Pocket Biotech Industry Primer

Yali Friedman, Ph.D.

LOGOS PRESS

Pocket Biotech
Industry Primer
by Yali Friedman, Ph.D.

Published in The United States of America
by
Logos Press, **Washington, DC**
WWW.LOGOS-PRESS.COM
INFO@LOGOS-PRESS.COM

Logos Press is an imprint of thinkBiotech LLC
Copyright © 2008, Yali Friedman, Ph.D.

10 9 8 7 6 5 4 3 2 1

ISBN-13:   978-1-934899-01-4

Contents

Figures and Tables

Chapter 1

Introduction

It is absolutely essential to recognize that success
comes at the end of failure after failure after failure
… If it were easy, 500 people would have already
done it.
Alejandro Zaffaroni

B iotechnology inventions and products are changing
paradigms in healthcare, agriculture, and industrial
processes. Great opportunities exist for those who
have the technologies, skills, and perseverance to bring new
biotechnology products to market. These opportunities stem
from the disruptive effects of biotechnology on existing
markets (and the ability to create new markets), but they are
tempered by a unique set of scientific, regulatory, political,
economic, social, and commercial influences. Understanding the dynamic and linked contributions of the interdisci-
plinary array of factors affecting the commercialization of
biotechnology is essential to operate in the biotechnology
industry.

The biotechnology industry is not defined by a set of
products or services, but by a set of enabling technologies.
Whereas the literal definition of biotechnology encompasses
everything from traditional agriculture to soap-making,
modern definitions describe applications relying on more
complex and sophisticated techniques such as genetic engineering and other forms of directed modification of living
things. This book defines biotechnology as the application
of molecular biology for useful purposes. This distinction is
important, because whereas inclusion of traditional activi-

Table 1.1 What is biotechnology?

Product / Service	Description
Biodegradable plastics	Reduce environmental impact of consumer goods
Diagnostic tests	Determine human predispositions to disease
DNA analysis	Determine paternity and assist in forensics
Genetic testing	Assist in traditional plant and animal breeding
Genetically modified crops	Improve yields and nutritional properties
Industrial enzymes	Improve efficiency and reduce environmental impact of industrial processes
Therapeutics	Treat and cure diseases

ties describes processes with established markets and mature technologies, the focus on modern techniques reflects the innovative and revolutionary possibilities of molecular biology: manipulating living organisms and parts of living organisms to capitalize on scientific discoveries, to improve upon existing solutions, or to serve new markets.

Biotechnology has applications in health, agriculture and farming, environmental remediation, and industrial processes. Within the diversity of biotechnology applications, there are two basic modes of development: products and services. Certain drugs, such as those produced in bacteria, yeast, and mammalian cells, are examples of biotechnology products (the distinction between biotechnology-derived and traditional pharmaceutical drugs is discussed in greater detail in Chapter 4). Drugs, and biotechnology research tools that are sold to pharmaceutical and other biotechnology firms, are also examples of products. Services can be sold to research firms or to companies further down value-chains for downstream application. Genetic testing is an example of a biotechnology service and is used to determine parentage, to resolve identity issues in criminal cases, and to screen for predispositions to disease.

The possible applications of biotechnology are defined by current scientific knowledge and abilities, and by the capacity of companies to develop marketable solutions from current knowledge or through additional research. The commercialization of biotechnology applications is further promoted and limited by numerous legal and regulatory factors. Patents serve both as a barrier to entry and an incentive for development. Changes in patent law can have profound implications on the ability of biotechnology firms to operate profitably and to obtain financing. Approval from bodies such as the Food and Drug Administration, the Department of Agriculture, and the Environmental Protection Agency is also required before many biotechnology products can be marketed or even tested.

Beyond these fundamental factors which define the possible applications of biotechnology, commercial factors also play an important role, as biotechnology ventures must ultimately be profitable. Whether structured as a for-profit company or a non-profit entity supported by donations or government grants, any biotechnology venture lacking an income stream cannot be sustained. Survival requires filling a need for which some party is willing to pay.

Chapter 2

The Development of Biotechnology

*In science the credit goes to the man who convinces
the world, not the man to whom the idea occurs first.*
Sir Francis Darwin

The modern biotechnology industry is built upon knowledge and techniques developed in the pharmaceutical industry, which employed biological extracts, dyes, and complex organic and chemical mixtures to produce drugs.

The emergence of the pharmaceutical industry is partially attributed to the development of aspirin, a drug that was developed by the German industrial chemist Felix Hoffman in 1897 and is still commonly used today. Many patients, including Hoffman's father, could not tolerate the stomach irritation associated with sodium salicylate, the standard anti-arthritis drug of the time. Armed with the knowledge that acidity associated with salicylates caused stomach discomfort, Hoffman sought a less-acidic formula and eventually produced acetyl-salicylic acid, or aspirin.

As medical knowledge advanced, a focus on symptom-based treatment of diseases replaced techniques such as blood-letting and led to research on the effects of medicines and the use of defined substances as drugs. The emergence of a rational basis for medicine supported research on human biology based on the belief that a better understanding of human biology would lead to better medicine. At the same time, improved knowledge of microorganisms related to human health led to an understanding of the causes of infectious diseases and allowed new treatment paradigms. Penicillin, for example, was identified as a potential anti-infective drug based on the observation of its ability to prevent the growth of bacteria in laboratory experiments.

The growth of the pharmaceutical industry paralleled advances in knowledge of general biology and advances in methods to study and manipulate biological systems. The emergence of refined tools permitted a more fundamental study of biology—molecular biology—focusing on the fundamental processes affecting biology. The discovery of the structure of DNA in 1953 was instrumental in developing an understanding of how genetically inherited characteristics are passed from generation to generation.

The first biotechnology companies were formed in the 1970s and 1980s. Knowledge of the molecular fundamentals of biology and development of tools to manipulate biological systems laid the foundation for the biotechnology industry, which employs the directed application of molecular biology for useful purposes. Biotechnology drug development not only uses methods and strategies different from traditional pharmaceutical development, it also produces different products. By selecting proteins such as insulin and erythropoietin, whose functions were already known, as their lead compounds, firms such as Amgen, Genentech, Chiron, and Genzyme employed a directed drug design strategy. In contrast with the chemical synthesis and biological extraction techniques that produced traditional pharmaceutical drugs, these early biotechnology companies used recombinant DNA techniques that enabled them to produce proteins as therapies (see Chapter 4 for more details).

KNOWLEDGE AND SKILLS

A brief history of selected Nobel Prize awards in the categories of Chemistry, and Physiology or Medicine provides a path to follow the scientific developments that spawned the biotechnology industry. Nobel Prizes are awarded for outstanding achievements and contributions and are internationally recognized as the most prestigious awards in the fields for which they are awarded. Because it can take some time for the significance of a discovery to emerge, many Nobel Prizes are awarded years after the actual discovery.

Frederick Sanger was awarded the Nobel Prize in Chem-

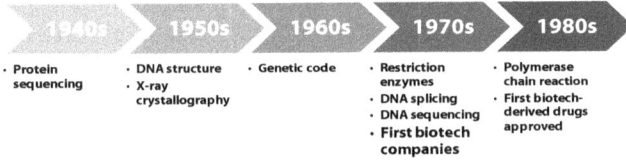

1940s	1950s	1960s	1970s	1980s
· Protein sequencing	· DNA structure · X-ray crystallography	· Genetic code	· Restriction enzymes · DNA splicing · DNA sequencing · **First biotech companies**	· Polymerase chain reaction · First biotech-derived drugs approved

Figure 2.1 *Knowledge and skills enabling biotechnology*

istry in 1958 for his determination of the protein sequence of insulin. Sanger, who began his mission in 1943, developed numerous techniques to directly sequence proteins, which enabled scientists to better understand these biological molecules. Knowledge of the sequence of human insulin enabled Genentech to develop recombinant human insulin—the first biotechnology drug—in 1982.

Between 1950 and 1956, Herbert Hauptman and Jerome Karle laid the foundations for the development of X-ray methods to determine the structure of crystallized molecules. They shared the 1985 Nobel Prize in Chemistry for their work. X-ray crystallography determines a molecule's three-dimensional structure by analyzing the X-ray diffraction patterns of crystals of the molecule. The complexity of organic molecules such as DNA and proteins meant that many structures were not known until the advent of X-ray crystallography. X-ray crystallography aided discovery of the structure of DNA in 1953, a significant advance in molecular biology that set the stage for modern biotechnology. James Watson, Francis Crick, and Maurice Wilkins shared the 1962 Nobel Prize in Physiology or Medicine for their work in discovering the structure of DNA. This discovery enabled elucidation of the mechanisms for control of gene expression and hereditary transfer of genetic information.

Following the discovery of the structure of DNA, the need to explain its role in cellular functions remained. Robert Holley, Har Gobind Khorana, and Marshall Nirenberg shared the 1968 Nobel Prize in Physiology or Medicine for their contributions in deciphering the genetic code, the language by which information is contained in DNA, and for elucidating how this information is translated by cells.

Werner Arber, Dan Nathans, and Hamilton Smith shared the Nobel Prize in Physiology or Medicine in 1978 for the discovery of restriction enzymes and their application to problems of molecular genetics. It was the pioneering work of these three scientists that enabled development of the DNA manipulation techniques that permitted Stanley Cohen and Herbert Boyer to develop methods for splicing DNA from different sources, often referred to as recombinant DNA (rDNA) technology.

The 1980 Nobel Prize in Chemistry was awarded to Paul Berg, Walter Gilbert, and Frederick Sanger. Berg was recognized for his "fundamental studies of the biochemistry of nucleic acids, with particular regard to recombinant-DNA," and Gilbert and Sanger for their "contributions concerning the determination of base sequences in nucleic acids." The ability to determine the sequence of DNA was central to the Human Genome Project and is a key element in biotechnology research and development.

Kary Mullis and Michael Smith shared the 1993 Nobel Prize in Chemistry for their respective development of the polymerase chain reaction (PCR), and site-directed mutagenesis. Mullis' PCR permits the specific production of copies of a specific DNA segment, even in the presence of a complex mixture of DNA. This technique has applications in forensics, paternity and heritage testing, medical diagnostics, archaeology and anthropology. Application of PCR and site-directed mutagenesis permits the directed modification of genetic sequences, effectively reprogramming genes.

APPLICATION

The significant scientific developments described above set the stage for the biotechnology industry. Understanding the role of DNA in programming the abilities of individual cells, combined with knowledge of how information is encoded in DNA, the mechanisms by which cells use this information, and the development of molecular biology techniques to manipulate DNA, gave rise to modern biotechnology.

In 1973, Stanley Cohen at Stanford University and Herbert Boyer at the University of California at San Francisco developed methods to splice genes and express foreign proteins in bacteria. This made it possible to deliberately make defined changes to biological systems, permitting the directed modification of microbes and cell cultures to produce desired products. Boyer and venture capitalist Robert Swanson formed Genentech in 1976, a defining event in modern biotechnology. Genentech, one of the first biotechnology companies, aimed to commercialize gene splicing technology by initially producing recombinant human insulin in bacteria to treat diabetes.

Prior to 1976, drugs were either chemically synthesized or extracted from living sources. Before bacterial production, insulin was commonly extracted from pig pancreas and required the sacrifice of 50 animals to produce sufficient insulin for a single person for one year. The advent of gene splicing introduced new possibilities, facilitating drug development without screening libraries of chemicals and biological extracts, and enabled scientists to select proteins whose function was already known as lead compounds.

Following proof-of-principle production of a neurotransmitter, Genentech produced recombinant human insulin in bacteria in 1978, later to become the first recombinant DNA drug approved by the Food and Drug Administration.

In 1980, prior to FDA approval of its recombinant human insulin, Genentech capitalized on positive market sentiment towards biotechnology and raised $35 million in an initial public stock offering. Without the resources to fully develop and commercialize recombinant human insulin as a drug, Genentech had licensed manufacturing and distribution rights to Eli Lilly, the dominant supplier of beef and pig insulin. Aiming to independently develop and commercialize a drug, Genentech became the first biotechnology company to market its own biopharmaceutical product in 1985 when it used gene splicing to produce human growth hormone, a drug previously available only by harvesting pituitary glands from deceased human organ donors. Since then, Genentech has produced many additional products, was bought by Roche Pharmaceuticals, and was subsequently resold on the public markets.

Genentech focused on one of the first core technologies defining the biotechnology industry, but is not the first biotechnology company. That status belongs to Cetus. Cetus was founded in Berkeley, CA, in 1971 and initially focused on using automated methods to screen for microorganisms with industrial applications. Despite developing the Nobel Prize-winning polymerase chain reaction technology, the company was not able to maintain independence, and was acquired by Chiron in 1991.

COMMERCIALIZATION

The history of Genentech serves as a paradigm for biotechnology product development and corporate growth. Genentech was founded to exploit a novel scientific innovation. Without sufficient resources to fully develop and commercialize its first product, Genentech licensed these rights to a larger partner. Tapping revenues from early products enabled Genentech to develop sufficient bulk to fully research, develop, and commercialize its own products.

The means and motivation must exist in order to develop a biotechnology product. The motivating factor can be as simple as consumer demand, permitting a company to derive revenues from sales. Alternatively, if a technology is sufficiently appealing, the potential to create new markets can motivate development. Conversely, public resistance to biotechnology products, such as opposition to genetically modified crops, can exert a negative influence on the marketability of a product. Whether a company is compensated directly from sales, government grants, or awards, or is compensated indirectly from tax credits, there must be some motivation to support development.

Legal and regulatory pressures can promote or discourage development. Long development times and the relative ease of reverse-engineering necessitate intellectual property protection for biotechnology products. Patents grant the right to exclude others from practicing an invention, providing an incentive for patent holders or licensees to develop patented applications by preventing competitors from capitalizing on their research and development investments. For this reason, many biotechnology firms form around patented scientific methods or proprietary knowledge that create a barrier to competitors and a source of revenue through licensing of partially- or fully-developed products and technologies.

A characteristic distinguishing biotechnology (and pharmaceutical) products from those of many other industries is the requirement for rigorous and lengthy assessments to verify the safety and, in the case of drugs, efficacy, of products prior

Box

Amgen: Capitalizing on innovation

Amgen was founded to capitalize on expanding opportunities in biotechnology. Founding CEO George Rathmann willingly left legacy pharmaceutical company Abbott for the more open and free environment of a biotechnology start-up. Today Amgen leads the biotechnology industry with revenues in excess of $14 billion and more than 20,000 employees.

1980: Amgen formed as Applied Molecular Genetics by a group of scientists and venture capitalists with a $19 million private-equity placement from venture capital firms and two major corporations
Employees: 3

1981: Amgen begins operations in Thousand Oaks, CA
Employees: 31

1983: Amgen isolates gene for human erythropoietin (EPO), later to become Epogen
Employees: 124

1985: Amgen sells Johnson & Johnson partial rights to EPO while still in development
Research team isolates gene for granulocyte colony-stimulating factor (G-CSF), later to become Neupogen
Employees: 196

1987: Amgen receives first patent for Epogen
Employees: 344

1989: Amgen receives first patent for Neupogen
FDA approves Epogen as orphan drug
Employees: 667

1991: FDA approves Neupogen
Employees: 1,723

1992: Amgen sales surpass $1 billion
Employees: 2,335

1996: Amgen sales surpass $2 billion
Employees: 4,646

1999: Amgen sales surpass $3 billion
Employees: 6,342

to being able to market them. Companies and financiers are therefore often unwilling to commit resources for development of drugs and other products for which the regulatory path is uncertain.

In addition to limiting development, government regulations can also motivate development. The Orphan Drug Act is an example of an incentive for drug development; tax credits and market exclusivity are granted to companies developing drugs for small populations that meet specific criteria.

Biotechnology development is fueled by innovation. The importance of specialized knowledge means that entrepreneurship by accomplished scientists is common in the genesis of biotechnology companies. The significant risk of product development failure compels biotechnology companies to focus on research and development until marketable products emerge. Patents and other barriers to entry are essential to prevent late-entering competitors from capitalizing on the efforts of pioneers.

INDUSTRY TRENDS

Many of the companies founded in the 1970s and 1980s sought to become fully vertically integrated drug developers, incorporating processes from drug discovery and development through production and sales. The prototypical company of this era aimed to develop treatments for unmet disease conditions and used the financing power of favorable public markets to fund expensive drug development efforts. Companies such as Genentech and Amgen were successful enough to achieve independence, but when market support for biotechnology disappeared, many companies had to reformulate their business models, merge, or liquidate.

Two impediments that prevented many of these early biotechnology companies from achieving vertical integration were the limited amount of available funding, which could not support the number of high-burn companies being founded, and the lack of experienced managers. The number of biotechnology companies aiming to become fully integrated diluted the amount of funding available at the time, limiting the support that each company could attain. Additionally, in order to develop vertically integrated companies, young startups needed managers with broad expertise from product development to commercialization. The only potential source for people with these skills was the pharmaceutical industry. Unfortunately, the pharmaceutical industry had divided the drug discovery and commercialization process into separate divisions managed by specialists, so no suitable managers existed. Furthermore, because biotechnology companies were seen as competitors, established pharmaceutical companies had little incentive for collaboration. By the late 1980s, pharmaceutical company sentiment towards biotechnology partnerships softened as pharmaceutical companies found themselves unable to maintain their growth rates solely by their internal research programs.

The 1990s saw the emergence of platform and tool-based companies seeking to commercialize drug targets, services, and technologies that could be sold or licensed to other companies. Revenue streams emerged from partner licensing fees, royalties, and research contracts. Although revenues from tools and services can make a company profitable, there is always the risk that these offerings can become commodities or obsolete.

Recognizing that revenues from tools and services could fund product development efforts, hybrid business models emerged in the late 1990s and early 2000s, capitalizing on the stability of tool and service sales while still selling the promise of product development. In addition to licensing or selling research tools to others, they were also used internally for product development. In principle, hybrid companies could therefore enjoy stable revenues from licensing and sales agreements while attracting investors by selling the promise of product development. The time and energy that must be devoted to marketing and selling tool offerings and keeping them current can make product development slower for hybrids than for product-focused companies. This reduced pace is balanced by the stability granted by revenues derived from tools which permit hybrid companies to better weather unfavorable financing environments.

The "no research, development only" (NRDO) model gained favor in the wake of the biotechnology bubble of 2000.

A derivation of the specialty pharmaceutical model of seeking additional markets for drugs already approved in one or more countries, the goal of NRDO firms is to acquire promising lead compounds and manage their clinical trials, at which point the drugs can be marketed in partnership with, or sold to, larger firms. NRDO firms were able to capitalize on the wealth of drug leads and managers that could be inexpensively acquired from firms struggling or liquidating as a result of unfavorable market conditions. A limitation of the NRDO model derives from the reality that many important discoveries in science emerge in the course of unrelated research. By not participating directly in research, NRDO firms are unable to realize the significant upside of tangential discoveries that emerge from research. A lack of internal drug development talent also challenges managers to obtain skilled guidance, often from paid consultants or contract research laboratories rather than internal experts, to assess the quality of potential product acquisitions.

Another recent trend is the move toward larger-scale projects. The ability to automate procedures such as DNA sequencing, microarray analysis, and drug screening make it possible to perform research at an unprecedented scale. Data mining and massive bioinformatics projects have also formed the core of companies. This shift in scale demonstrates a very important change in the way research is conducted. The ability to perform large-scale experiments requires reliability and automation, attributes not often found in basic scientific discoveries and methods. DNA sequencing, a procedure that can now be fully automated, once required days of manual labor. Just as computers have advanced knowledge in other disciplines with their ability to process information and reliably and repeatedly perform tasks, the ability to automate biotechnology experiments will lead to greater discoveries at lower costs.

Chapter 3

Introduction to Molecular Biology

Everything should be as simple as possible, but not simpler.
Albert Einstein

Biotechnology research seeks to develop applications of molecular biology. Many sources use analogies to recipe books or blueprints to explain the role of DNA and genes in molecular biology. Ultimately, these analogies obscure the importance of topics such as regulation of gene expression, which is of fundamental importance in understanding molecular biology. When applying one's knowledge of biotechnology fundamentals, most metaphors fail. It is only by understanding molecular biology and biotechnology applications that one can appreciate the applications and limitations of techniques used in molecular biology.

This chapter presents a brief, metaphor-free, introduction to molecular biology. Subsequent chapters describe the tools, techniques, and applications of biotechnology and provide greater details on the potential and limitations of molecular biology.

INFORMATION FLOW IN MOLECULAR BIOLOGY

In order to understand the basis of most biotechnology applications, it is necessary to first understand the process by which information in genes leads to the formation of structural and functional proteins.

Proteins serve structural and functional roles that give individual cells—and by extension whole organisms—specific structures and functional characteristics. When many people think of proteins, they think of foods such as meat and beans.

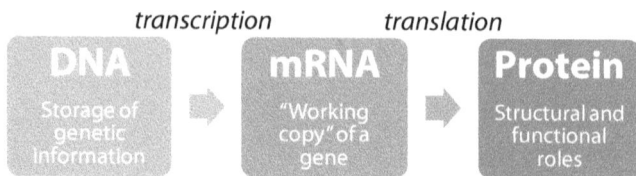

transcription   *translation*

DNA	mRNA	Protein
Storage of genetic information	"Working copy" of a gene	Structural and functional roles

Figure 3.1 *Simplified model of information flow in molecular biology*

While animal muscle and plant seeds are excellent sources of dietary protein, proteins play a central role in all cell types and perform functional and structural roles (see Table 3.1). Examples of structural proteins include keratin, which makes skin waterproof, and myosin, which interacts with other proteins in muscles to make them flex.

DNA contains information that describes the construction of proteins. The process of protein synthesis is as follows:

1. DNA contains the information to produce proteins.
2. Information encoded in DNA is *transcribed* into a molecule called messenger RNA (mRNA)—effectively a "working copy" of the DNA sequence of a given gene.
3. mRNA is *translated* into proteins by the protein synthesis machinery, the composition of the resulting protein corresponding to the original DNA instructions.

This basic mechanism is conserved in all life forms, from bacteria to humans. The implication of this common process that converts information in DNA into functional proteins is that similar techniques can be used to investigate and manipulate all biological systems. Furthermore, it is possible to make human therapeutic proteins, for example, in organisms as distantly related as bacteria.

Understanding the roles of DNA, RNA, and protein and their relationships to each other is essential to understanding molecular biology. While there are some specific exceptions (e.g., retroviruses and prions) to the order and direction of information flow shown in Figure 3.1, these examples still fit within the general framework, and the majority of biological systems use the framework as presented.

DNA: STORING AND RELAYING INFORMATION

Deoxyribonucleic acid (DNA) is the primary source of genetic information in cells. Humans, plants, animals, and bacteria all contain DNA. DNA is physically passed from generation to generation, bestowing certain traits of parents to their children. The reason why children have physical characteristics from each of their parents—a child may have their mother's eye color and father's hair color—is because they received half their DNA from each parent.

Each of our cells (with a few exceptions like red blood cells, eggs, and sperm) contain all the DNA required to code our genetic features. Individual regions of DNA that confer traits are called genes. Information in genes is relayed to the protein synthesis machinery within cells where it dictates the production of proteins. The word "genome" refers to all the DNA in an organism. The human genome contains approximately 30,000 genes arrayed on 46 long stretches of DNA called chromosomes.

DNA is essentially composed of two intertwined strands that form a double helix. The two strands of DNA are said to be complementary because the sequence of one strand indicates the sequence of the opposite strand, like a photograph and its negative. Each strand is physically composed of four different

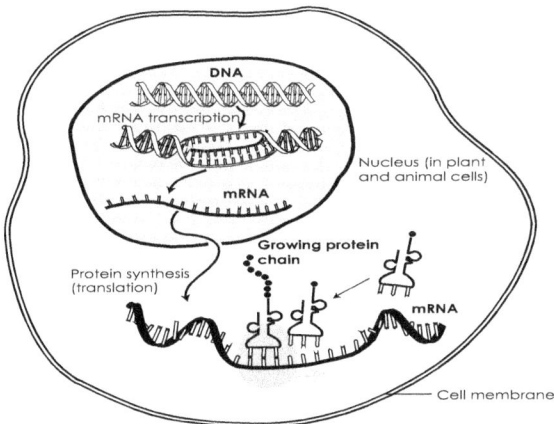

Figure 3.2 *General scheme of gene expression*
Modified from National Human Genome Research Institute

chemical units called nucleotides, the sequence of which encodes the genetic information. These four chemical units, adenine, cytosine, guanine, and thymine, are often abbreviated as A, C, G, and T, respectively. Just as the English language can be expressed in twenty-six letters, the genetic code is expressed in these four chemical units. A DNA "sequence" refers to the specific order of A's, C's, G's, and T's in a stretch of DNA.

There are two essential elements of genes: coding and regulatory elements. The coding elements of genes are first transcribed as mRNA, which is then translated into protein. The chemical sequence of A's, C's, G's, and T's in the coding region of a gene determines the composition and structure of the resulting protein and, by extension, its function. Regulatory elements affect the rate at which genes are transcribed and translated, and may be interspersed within the coding sequence or

---

Box

Human chromosomes and genetic trait inheritance

The human genome is composed of chromosomes. We get 23 chromosomes from our mother and 23 chromosomes from our father, constituting 23 pairs. While 22 of the 23 chromosome pairs are similar in both men and women, the 23rd pair is quite different and determines the sex of an individual. For the 23rd pair of chromosomes, women have two X-chromosomes while men have one X- and one Y-chromosome. Because X-chromosomes contain more DNA than Y-chromosomes, they are physically larger than Y-chromosomes. Having too many or too few chromosomes can affect gene regulation and cause diseases. Down's Syndrome, for example, occurs in individuals with three copies of chromosome 21.

The roles of X- and Y-chromosomes are important in understanding sex-linked diseases. Women do not have Y-chromosomes, so diseases that are caused by defective genes on the Y-chromosome can only occur in men. Additionally, men only have one X-chromosome, so mutations in genes on the X-chromosome are more likely to affect males, because the second X-chromosome in women can sometimes compensate for mutations on the first. Color blindness, caused by a mutation on the X-chromosome, is more common in men than women for this reason.

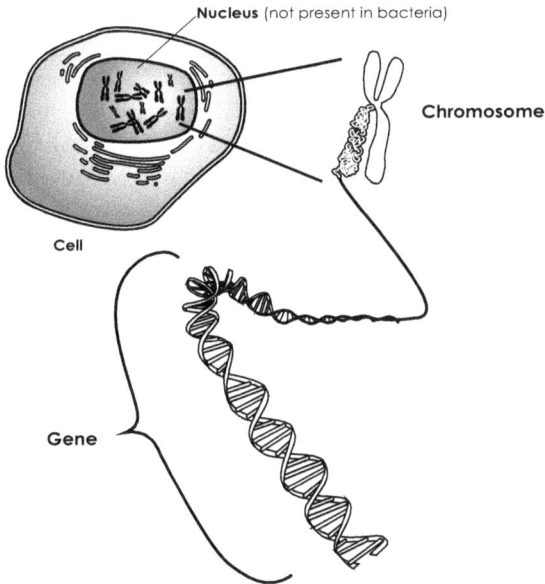

Figure 3.3 *DNA: Chromosomes and genes*
Modified from National Human Genome Research Institute

outside of it. Regulatory elements also control the cell types within which specific genes are activated, and the timing and magnitude of gene expression. Gene regulation thereby allows individual proteins to be expressed only in certain cells at specific times and at specific rates.

Proper regulation of gene expression—the production of gene products—is essential. Under- or over-expression of genes can have deleterious effects. For example, many forms of cancer are caused by mis-regulation of gene expression that results in uncontrolled cell division. A potential solution for diseases resulting from low expression of genes is to use gene therapy to introduce affected genes or regulatory elements to spur additional production. One of the challenges of gene therapy is developing methods to regulate the expression of genes that are introduced into cells and ensure that they are not over-expressed. A solution for diseases caused by over-expressed genes is RNA interference. This procedure prevents translation of mRNA, inhibiting protein production. RNA interference is described in further detail in Chapter 6.

MRNA: THE MESSENGER

Messenger RNA (mRNA) is used to relay information from genes in DNA to the protein synthesis machinery. An additional feature of mRNA is that it can be destroyed once sufficient protein is produced, permitting an extra level of control of gene expression. RNA is also present in forms other than mRNA, some of which are described later in this chapter.

It is possible to affect expression of genes by targeting their mRNA with antisense RNA or DNA—nucleic acids which can bind the mRNA. Unlike DNA, which is usually double-stranded, mRNA is single stranded. Nucleic acids (DNA or RNA) containing a sequence that can bind to a given mRNA will prevent translation by the protein synthesis machinery, inhibiting gene expression. The Flavr Savr tomato, a tomato engineered to have a long shelf life, was produced by introducing antisense RNA corresponding to mRNA for an enzyme involved in fruit spoilage. Inhibiting expression of this gene delays spoilage. In 1998 the FDA approved Isis Pharmaceuticals' Vitravene, the first antisense drug, to treat cytomegalovirus-induced retinitis.

TRANSLATION: MAKING PROTEINS

Just as DNA and RNA are composed of linked nucleotides, proteins are comprised of chains of amino acid units. When mRNA is translated to produce a protein, the protein-synthesis machinery "reads" the nucleotides three at a time, assembling amino acid chains that correspond to the mRNA sequence. The basic elements of the protein-synthesis machinery are tRNA, a form of RNA that *transfers* amino acids to the protein-synthesis machinery in a way that enables them to be linked together, and ribosomes, which help form the chemical bonds that attach amino acids in a protein chain.

The three-nucleotide sequence elements on mRNA that code for individual amino acids are called codons. These are matched by anti-codons on tRNA to ensure that the appropriate amino acid is aligned with a given mRNA sequence. The 64 possible combinations of A, C, G, and T at each codon code for

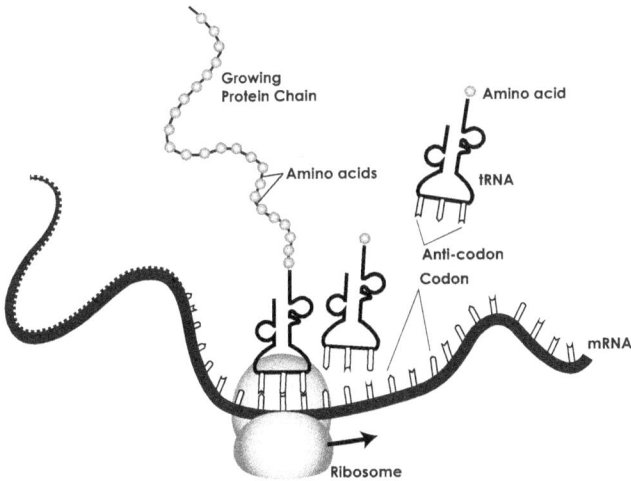

Figure 3.4 *Protein translation*
Modified from National Human Genome Research Institute

only 20 different amino acids. This redundancy in the genetic code, permitting multiple codons to specify common amino acids, is considered a form of protection against DNA mutations and has applications in identifying foreign DNA from sources such as viruses which may use different "dialects" of the genetic code.

The chemical characteristics of amino acids in a protein cause it to fold into a defined 3-dimensional structure. That determines the protein's function. Because the DNA sequence of a gene dictates the sequence of amino acids in a protein, and the sequence of these acids in a protein determines its structure, one can deduce a protein sequence, and potentially its structure and function, from the gene sequence encoding it.

PROTEINS AND ENZYMES

Proteins, the workhorses of cells, are responsible for the majority of structural features and functional characteristics in cells. Enzymes are proteins that perform functional roles as part of the cellular process. Different types of cells get their characteristics by expressing a specific array of genes, resulting in production of a complement of proteins that give each cell

Table 3.1 *Examples of protein and enzyme functions*

Enzyme	Function
Amylase	Breaks down starches and other complex carbohydrates into basic sugars
Cellulase	Breaks down cellulose, found in the cell walls of plants
Lipase	Breaks down fats
Protease	Breaks down proteins

Protein	Function
Collagen	Main protein in connective tissue; structural roles in skin, cartilage, teeth, bone, and other tissues
Keratin	Makes skin waterproof and contributes to strength and flexibility
Myosin	Muscle contraction

type its unique characteristics. Pancreatic cells, for example, produce the protein insulin to regulate blood sugar levels; neurons produce neurotransmitters essential for brain function; and hemoglobin is made in blood cells, enabling them to carry oxygen. Examples of enzymes include proteases that break down proteins or enable digestion of food, and polymerases that assemble DNA and RNA. Some genes are expressed only in certain cell types whereas others are widely expressed. Examples of widely-expressed genes include those encoding proteins and enzymes involved in general cellular activities such as DNA replication, mRNA translation, protein synthesis, energy production and maintenance of structural integrity.

Production of inappropriate proteins in cell types and mis-regulation of protein expression are at the root of many diseases. As mentioned above, many cancers result from mis-regulation of gene expression that causes uncontrolled cell division.

Molecular biologists can transfer genes from humans and other animals into bacteria, yeast, and other organisms to confer the ability to produce specific proteins that may be extracted for therapeutic use. For example, Genentech produced its first drug by introducing the gene for human insulin into bacteria and extracted the resulting protein to produce a treatment for human diabetes. Genes can also be transferred from one or-

Table 3.2 Selected *RNA types*

RNA type	Function
mRNA	Messenger RNA. Contains a working copy of a gene sequence and is read by the protein synthesis machinery to produce proteins.
tRNA	Transfer RNA. Transfers amino acids to the protein synthesis machinery to produce proteins.
rRNA	Ribosomal RNA. Part of the protein synthesis machinery. Also useful for determining evolutionary similarity between organisms.
aRNA	Antisense RNA. Used for gene regulation.
siRNA	Small Interfering RNA. Used for gene regulation.
snRNA	Small Nuclear RNA. Used to edit mRNA, regulate gene expression, and maintain chromosome tips (telomeres).

ganism to another to confer new attributes. Pesticide-resistant crops have been produced by incorporating naturally-occuring pesticidal proteins into plants. Bacteria have also been modified to perform roles such as decomposing oil spills by adding genes encoding proteins with the ability to break down components of oil. Additional examples are described in Chapter 6.

OTHER FORMS OF RNA

Traditional molecular biology held that the primary role of RNA in cells was largely limited to housekeeping functions such as transferring information from DNA to the protein synthesis machinery (mRNA), transporting amino acids to be assembled into proteins (tRNA), and translating mRNA into protein (rRNA).

Sidney Altman and Thomas Cech shared the 1989 Nobel Prize in Chemistry for their discovery of catalytic properties of RNA. The ability to catalyze (increase the rate of) biochemical reactions had previously been thought to only exist in proteins. Altman and Cech found a role for RNA in the splicing of mRNAs, ultimately making it possible for a single gene to give rise to several different proteins. The significance of Altman and Cech's discovery was expanded more than a decade after they received the Nobel Prize. Following sequencing of the human genome it was discovered that the human genome contained only a fraction of the genes previously thought necessary to produce the complete set of proteins comprising human biology. The ability of this small set of genes to produce the full complement of human proteins could largely be explained through mRNA splicing.

More recently myriad forms of RNA have been discovered, and diverse roles for RNA have also been elucidated (see Table 3.2). These discoveries indicate that controlling cellular activities is more complex than previously thought, suggesting that there are also more opportunities to influence cellular activities.

THE BIG PICTURE

Genes interact with the environment and with each other to confer traits. While the presence or absence of a gene can potentially confer a given trait, environmental factors also play a role. Our physical characteristics are a combination of genetic and environmental factors. A child with a hypothetical *tallness* gene, for instance, would not necessarily grow taller than a child without the gene; the child with the *tallness* gene would also require adequate nutrition to fuel the extra growth (and the effect of the *tallness* gene may be limited or enhanced by the action of other genes). Rather than thinking of genes as determinants of physical characteristics, they should be regarded as potentials or predispositions for characteristics.

The ability to modify characteristics of cells is similarly limited by biological and physical constraints. Since some cells are rapidly replaced, induced changes will be quickly lost. Other cells are dormant, precluding their potential to express modifications.

Furthermore, biology is complicated. In fields such as industrial chemistry or engineering, applications are developed from well-characterized principles. With biotechnology on the leading edge of molecular biology research, it can be difficult or impossible to foretell the outcomes of manipulations and they can have unforeseen consequences. Because it is not possible to fully predict the outcome of these procedures, scientists must perform experiments, take observations, refine theories, and finally develop functional applications. This is why biotechnology research is so complex, time consuming, and fraught with unforeseen setbacks and disappointments.

Chapter 4

Drug Development

*It is only by the means of the sciences of life that the
quality of life can be radically changed.*
Aldous Huxley

D rugs are substances that affect the functions of living things and are administered to treat, prevent, or
cure unwanted diseases and symptoms. The United
States Food and Drug Administration (FDA) regulates drug
marketing, requiring manufacturers to prove their products to
be safe, effective, and appropriately labeled. The drug development process identifies drug candidates and subjects them to
increasingly stringent tests to assess their safety and efficacy.
Drug development is paradigmatic of the general process by
which biotechnology products are developed, with one important difference: non-therapeutic products are not subject to the
same regulatory pressures to gain marketing approval.

Scientists start with simple, defined, model systems that
enable them to identify potential drugs. These potential drugs
are then tested in increasingly complex and real-world situations to prove their efficacy. It is important to test for as many
contingencies as possible. Something that works well in a simple model system may fail in real-world use due to any number
of unforeseen circumstances.

The description of drug development in this section is
presented as a model for biotechnology product development.
Producing and selling drugs consists of three basic stages: discovery, development, and commercialization. Less than 1 percent of early candidate compounds make it through the drug
development process.

Discovery-stage research produces lead compounds that

must pass tests to predict their toxicity, to determine if they can be effectively administered, and to project the likelihood of recouping development costs (see Figure 4.4).

The development process builds on observations and products from discovery-stage research. Formulations are developed to optimize drug administration, and pre-clinical and clinical trials are employed to test the safety and efficacy of drugs in humans. In addition to testing the physical properties of a drug, manufacturing processes which can consistently produce doses of equivalent purity and efficacy over a period of time must be developed and tested.

BIOTECHNOLOGY VS. PHARMACEUTICAL DRUG DEVELOPMENT

Traditional pharmaceutical drugs differ from biotechnology-derived drugs in the methods by which they are discovered and manufactured. As a result, the resulting drugs have markedly different characteristics.

To distinguish traditional pharmaceutical from biotechnology drug development, consider the traditional pharmaceutical and biotechnology forms of therapeutic insulin. Prior to Genentech's production of recombinant human insulin, pharmaceutical companies extracted insulin from the pancreas of pigs, cows, and horses. Glands from fifty pigs were needed to produce sufficient insulin to treat a single person for one year. Insulin from these sources was subject to disease transmission, shortages, and reactions with the human immune system. Genentech produced recombinant human insulin, the first biotechnology drug, by synthesizing it in bacteria. Bacterial fermentation allowed for greater production capacity, avoidance of immune system reactions typical of non-human forms of insulin, and elimination of the threat of transmission of animal diseases. This example illustrates how the fundamentals-based approach to product development employed by biotechnology firms permits the development of solutions not attainable by traditional pharmaceutical development.

Traditional pharmaceutical drug discovery was based on trial-and-error screening of synthetic compounds and directed selection of biological extracts that can affect model systems.

Small-Molecule Drug

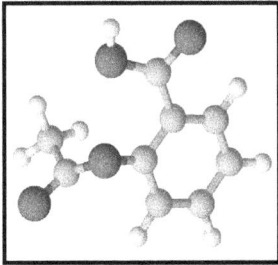

Aspirin
23 atoms

Biologic Drug

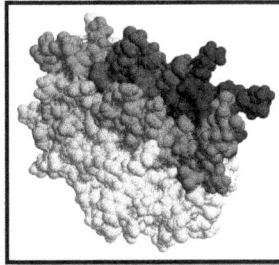

Erythropoietin
1297 atoms

Figure 4.1 *Small-molecule and biologic drugs*

The emphasis of research was to understand biological systems in order to find potential drug targets. Compounds and extracts that interacted with these targets were then selected for further study to see if they could be used as drugs.

The molecular biology techniques used by biotechnology firms differ from traditional pharmaceutical development because they permit a finer-scale analysis of biological systems and the directed design of biological compounds as drug candidates. Traditional pharmaceutical development was limited to chemical synthesis and biological extracts. The reason for this limitation was that traditional pharmaceutical development originated before the advent of molecular biology techniques, which enable the directed design and production of biological molecules.

Drugs produced by traditional pharmaceutical means tend to be small molecules that are orally doseable as tablets, capsules, or liquids (see Figure 4.1). Following absorption in the gastrointestinal tract these drugs travel throughout the body in the bloodstream, and can often be mass-produced for a relatively low cost.

While biotechnology research techniques do enable new possibilities, they have not rendered pharmaceutical research techniques obsolete. Traditional pharmaceutical research is still practised because of available expertise, the abundance of chemical and biological-extract libraries, and the strength of

2006 Biotechnology drug sales by category

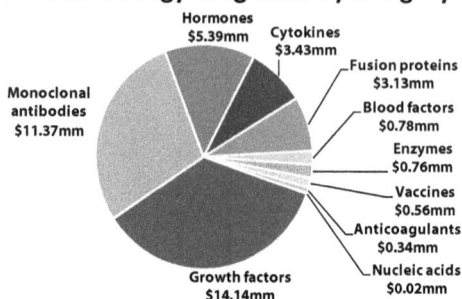

Figure 4.2 *Biotechnology drug categories*

techniques for target selection and optimization.

The majority of biotechnology drugs have been proteins, such as growth factors, monoclonal antibodies, hormones, and cytokines (see Figure 4.2). Other categories include nucleic acids and vaccines. The ability to design, modify, and synthesize biological compounds means that many biotechnology drugs are larger and more complex than traditional pharmaceutical drugs.

Drug delivery is an issue for biotechnology-derived drugs because proteins and other biotechnology drugs such as nucleic acids are less likely to survive the acidic conditions in the stomach and are generally unable to pass through the intestinal lining and travel the bloodstream to their therapeutic target. Biotechnology drug delivery techniques include injection, skin patches, and inhalation. Some of these alternative delivery systems allow for more precise tissue targeting and improved dosage control, presenting new opportunities.

Because the post-discovery activities for biotechnology-derived and traditional pharmaceutical drugs are relatively similar, modern pharmaceutical firms are also able to develop lead compounds generated by biotechnology firms, and actively engage in biotechnology research themselves. Conversely, as the biotechnology industry has matured, some biotechnology companies have developed late-stage development and marketing abilities on par with small and medium-sized pharmaceutical companies. These shifts in the roles of pharmaceutical and

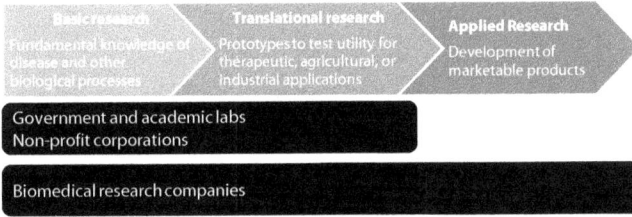

Figure 4.3 *Basic and applied research*

biotechnology firms have led to a blurring of the distinction between the two.

THE FIVE BASIC STEPS OF DRUG DEVELOPMENT

There are two fundamentally different types of research conducted in early-stage drug development: basic and applied research (see Figure 4.3). Basic research is directed at improving fundamental knowledge of biological systems, disease processes, and potential points of therapeutic intervention. Applied research utilizes this knowledge to identify and develop the therapeutic agents themselves. This distinction is important because different toolsets and mindsets are involved in these two types of research. Basic research generally does not directly produce drugs. Instead, it lays the foundations upon which drugs are developed and produced. Genomics, proteomics, molecular physiology, and other basic research areas contribute essential information for disease characterization and target identification. Applied research is required to enable further development. Translational research forms a bridge between basic and applied research, testing principles from basic research and generating "proof-of-principle" in preparation for applied research.

STEP 1: IDENTIFY A USEFUL DISEASE TARGET

A fundamental understanding of the science involved in a therapeutic problem is essential for drug development. Ideally, scientists start with an understanding of the molecular processes affecting conditions they wish to treat and test hypotheses about which drugs are likely to be effective for a condition.

Figure 4.4 *The process of drug development*

Model systems are employed because it is unethical and impractical to test uncharacterized candidate drugs on humans. These model systems generally progress in complexity through the stages of drug development. Early-stage model systems may be as simple as a set of molecules in a test tube (subject to knowledge of disease processes) and later-stage model systems may be live animals with diseases similar to human conditions (subject to availability of relevant disease models).

To work on a problem effectively, the tools must be well-defined. Poor model systems or a lack of precise measures can yield experimental results which are misleading or difficult to interpret. It is vitally important to have a fundamental appreciation of the problem to be solved and to apply the appropriate tools to assess potential solutions. Many biotechnology applications, such as gene therapy and RNA interference (described in Chapter 6) are challenged by poor availability of predictive models.

STEP 2: FIND AND REFINE A LEAD COMPOUND

Potential lead compounds typically originate from one of two sources: purified naturally occurring compounds or *de novo* design and synthesis of new compounds. As described earlier in this chapter, biotechnology drugs have different properties and capabilities than traditional pharmaceutical drugs. The traditional pharmaceutical method for drug discovery involves screening libraries of natural or synthetic compounds to find those that achieve the desired effect in model systems. The molecular biology techniques used in biotechnology drug development permit directed selection of natural compounds— usually biological molecules—and the design and synthesis of novel biological compounds as drug candidates.

As knowledge of biological systems improves, it becomes increasingly possible to refine screening methods. Knowledge of disease mechanisms can also help build better model systems through identification of appropriate therapeutic targets as well as other targets which might be a source of undesirable side effects.

An understanding of how biological systems operate at the molecular level, combined with the ability to express gene products in bacteria, yeast, and animal-derived cells, enables development of drugs based on specific disease requirements. Many traditional pharmaceutical drugs were discovered by screening libraries for effective leads, rather than starting with knowledge disease processes and working backwards to find a solution. Knowledge of the structure of a key molecule involved in a disease, such as the HIV protease that is integral to AIDS, enables *in silico* (computer model-based) techniques, using computers to select or design compounds likely to inhibit the enzyme.

Studying herbal remedies used by different cultures, indigenous peoples, and animals is also an excellent source for potential drug compounds. Some naturally occurring compounds, such as penicillin, are used as drugs based on their natural activities in biological systems. The antibiotic penicillin was identified as the factor that permits *Penicillium* mold to inhibit bacterial growth. Whereas penicillin is naturally produced by *Penicillium* as an antibiotic, the same therapeutic application it is used for in humans, other natural compounds, such as *botulinum* toxin, are used for novel purposes such as treating abnormal muscle contractions and in cosmetic applications.

Another method to discover new drugs is to examine the side effects of existing drugs. Minoxidil, now prescribed as a topical treatment for hair loss, was initially intended for the treatment of severe blood pressure. The curious side effect of stimulating hair growth led to its use as a treatment for balding. In another example, Viagra's potential for the treatment of erectile dysfunction was discovered in clinical trials for treatment of angina.

Once a potential drug that works in a model system is identified, it is time to study and refine its activity. This potential drug is called a lead compound. Aside from drug activity, factors such as a drug's shelf life at different temperatures, ease of large-scale manufacture, and lot-to-lot consistency must also be considered. In optimizing lead compounds, researchers aim to identify the elements that are essential for their activity and modify those elements to obtain optimal efficacy and/or safety properties.

STEP 3: TEST LEAD IN PRE-CLINICAL DEVELOPMENT

In pre-clinical development, lead compounds that emerge from the lead optimization process are subjected to a range of standardized animal, cellular, and biochemical tests designed to gauge their suitability and safety for human administration as well as to estimate the range of dose levels of the compound that will be utilized in subsequent human trials. Animal models are also used to provide preliminary assessments of the absorption, degradation, and potential toxicity of drugs. During pre-clinical development, these animal tests are conducted under much more tightly prescribed conditions, such as the industry-standard GLP (Good Laboratory Practices) procedures, which have stringent quality control and quality assurance oversight than is normally used in earlier stages. While it may be possible to produce a desired effect in a model system in a laboratory setting, real-world situations often present unforeseen obstacles. For example, many gene therapy techniques that work in cultured cells in laboratory settings fail when introduced into human beings.

Scientists use animals to test toxicity and attempt to cure animal versions of human diseases before proceeding to human trials. Because biotechnology enables biological changes that were previously impossible, it is not possible to predict all the implications. Many products that work well in laboratory tests fail in clinical settings for unforeseen or even improbable reasons; drugs may not be taken up properly by cells; they may be metabolized into inactive or toxic forms by the liver; they may interact with other parts of the body to produce undesired

Drug development stages

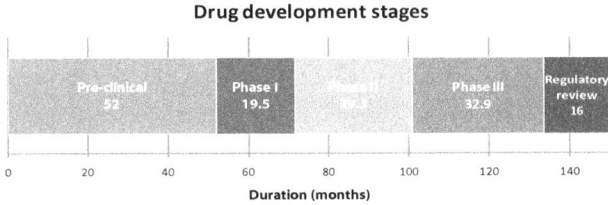

Figure 4.5 *Biotechnology drug development time*
Source: Dimasi, J.A., Grabowski, H.G. The cost of biopharmaceutical
   R&D: Is biotech different? *Managerial and decision economics*, 2007.
   28:469-479.

effects; or they may simply not be sufficiently active. This is one of the reasons why animals are so important in drug research.

While success in animal tests does not necessarily mean that a compound will work in humans, a compound that performs poorly in animals is unlikely to work well in humans. Ultimately, human testing is necessary for the safety and efficacy determinations required for FDA approval.

In addition to testing the drug candidate in animal models, another set of activities that takes place during pre-clinical development is development of methods for manufacturing and formulating the drug on a commercial scale. Unlike research costs, manufacturing costs recur over the life of a drug. Minimizing these recurring costs can significantly impact profits. This examination is also important for patent protection because it may lead to additional patent claims, potentially impeding the development of competing drugs.

STEP 4: CLINICAL TRIALS IN HUMANS

The clinical trial process investigates drugs for safety and efficacy in humans. There are four "phases" of clinical trials. Phases I through III demonstrate safety and efficacy prior to approval, and Phase IV monitors safety post-approval and tests new treatment indications. On average, one compound in a thousand will make it to clinical trials. Roughly 70 percent of drugs that complete clinical trials receive FDA approval.

While safety is the primary concern, a drug with detrimental side effects may be acceptable if there are no better treatments and the severity of disease warrants it. Most companies file for and receive patents for the commercial use of compounds during pre-clinical development. Much of a drug patent's life can therefore lapse during clinical trials and while waiting for regulatory review.

Every clinical trial in the United States must be monitored by an Institutional Review Board (IRB), an independent committee of physicians, statisticians, community advocates, and others that ensures that the risks are as low as possible and are worth any potential benefits, and that the rights of study subjects are protected.

PHASE I

While the purpose of clinical trials is to determine the safety and efficacy of a drug, the primary consideration is the safety of the participants. Phase I trials are designed to determine the safety of drugs. These trials involve a small number of healthy volunteers or affected patients who are given doses ranging from sub-clinical to potentially toxic.

To minimize risk to human subjects, all drugs must undergo extensive pre-clinical development to determine their effects on animals, and predicted effects in humans, prior to Phase I trials.

Beginning with human trials in Phase I, drugs must be produced under current good manufacturing practices (cGMP). To satisfy FDA cGMP guidelines, manufacturers must be able to demonstrate compliance with regard to facilities, raw materials handling, and manufacturing control and associated documentation (see *Manufacturing* in Chapter 5).

There are two basic types of Phase I trials. *First-in-man* studies are primarily concerned with establishing the safety of a compound. These studies start with an initial small dose that is given to a small group of participants. If no adverse effects are seen, escalating doses are given to new groups of participants. Dose limiting toxicity is observed when the dose is escalated to the point that dangerous side effects are seen. The other type of Phase I study is the *clinical pharmacology* study. The goal of this study is to determine the pharmacokinetics of a compound: how a drug is absorbed, distributed in the body, metabolized, and excreted.

Phase I testing ranges from one to three years on average. Historically, drugs in Phase I have a 10 percent chance of making it to market. If Phase I trials do not reveal unacceptable toxicity, a drug can proceed to Phase II testing. While failure in a phase I trial indicates that the tested form of a drug is unacceptable, success may still be possible by modifying a compound based on observed data.

Phase II

The emphasis in Phase I trials is on safety; Phase II trials introduce effectiveness. Phase II trials consist of small, well-controlled experiments to further evaluate a drug's safety, assess side effects, and establish dosage guidelines. Drugs are given to volunteers (usually between 100 and 300 patients) who actually suffer from the disease or condition being targeted by a drug.

This phase is where the minimum effective dose, maximum tolerable dose, and optimum dosage of drugs are established. Drug regimens are tested to see how often a drug must be administered; a drug may be effective if taken once a month, or may require administration several times a day. Statistical end points are established for drugs, representing the targeted favorable outcome of the study. The current standard of care for a medical condition can be used as a benchmark in setting the end point.

Phase II trials last an average of two years. If Phase II trials indicate effectiveness, a drug can proceed to Phase III trials. A drug that moves on to Phase III testing has an approximately 60 percent chance of being approved by the FDA. A properly designed and administered Phase II trial can help select dosage regimens and treatment indications that make Phase III trials faster and easier. Rushing this process may require repetition of Phase III trials or lead to outright failure.

Phase IIA / IIB

Because clinical trial phases can take years to complete and often have multiple objectives, drug developers have taken to sub-dividing the phases in order to express a sense of progression. This division is most prevalent in Phase II trials, where safety data from Phase I trials are confirmed and expanded, and dosage and administration profiles are established in preparation for Phase III trials.

While the terms Phase IIa and Phase IIb are not recognized by the FDA, they are used useful devices to convey a drug's position in the approval process to investors, analysts, and partners. As a general rule, Phase IIa trials tend to address expansion and confirmation of data from Phase I trials—absorption, metabolism, and pharmacodynamics—whereas Phase IIb trials resemble small-scale Phase III trials in their evaluation of safety and clinical efficacy in large populations.

PHASE III

Phase III testing is the largest and most expensive clinical trial phase, and is intended to verify the effectiveness of a drug for the condition it targets, based on statistical end points established in Phase II trials. Phase III trials also continue to build the safety profile of drugs and record possible side effects and adverse reactions resulting from long-term use.

Phase III trials are tightly controlled, preferably double-blind, studies usually with at least 1,000 patients. In double-blind studies, neither patients nor the individuals treating them know whether the active drug or an alternative such as a placebo is being administered. Relative to Phase I and Phase II trials, the larger and ideally more diverse populations used in Phase III trials are necessary to determine if certain types of patients develop side effects or do not respond to treatment.

Two successful Phase III trials are generally required to ensure the validity of the studies, although a single trial may suffice if the results are extremely strong. Phase III testing averages between three and four years.

STEP 5: OBTAIN APPROVAL; MARKET AND SELL DRUG

The FDA requires that drugs be approved prior to marketing. While safety is the primary concern, a drug with detrimental side effects may be acceptable if there are no better treatments and the severity of the disease warrants it.

Current estimates of development times for small-molecule drugs are 10-15 years with an estimated average cost of $802 million per approved drug.[1] It is worth noting that half of this cost is attributed to financing costs, reflecting the "opportunity-cost of capital" invested over the 10-15 year timeline. Roughly one-third of the expenditures are attributed to pre-clinical activities and the remaining two-thirds to clinical activities. The Tufts Center for the Study of Drug Development found that only five in five thousand small-molecule compounds that enter pre-clinical testing make it to human testing. Of these five, only one is approved.[2] These numbers were derived from examination of small molecule synthetic drugs, which are produced by traditional pharmaceutical techniques and differ from biologic drugs in several important ways. Estimates for the cost of biologic drug development are $1.2 billion, comprised of approximately $500 million in out-of-pocket expenses, and $700 million in capitalization costs.[3] Both these cost estimates include the cost of failed leads. Therefore, they do not predict the expenditures required to produce a single drug; they predict the investments required by successful and failed research projects that result in the development of a single drug.

Relative to small molecule drugs, the sample size to evaluate development times and costs for biologics is much smaller and subject to bias—the first biologics had shorter development times than later entrants—but initial indications are that development times and costs for biologics are similar to those for small molecules. It is important to note that any estimate of drug development time or cost is profoundly affected by context. First-in-class drugs, drugs serving new markets, or drugs serving pressing needs are likely to require smaller and fewer clinical trials than drugs with little differentiation from existing alternatives or those serving less-pressing needs, decreas-

1   DiMasi, J.A., Hansen, R.W., Grabowski, H.G. The price of innovation: new estimates of drug development costs. *Journal of Health Economics*, 2003. 22:151–185.

2   How new drugs move through the development and approval process. *Tufts Center for the Study of Drug Development*, November 1, 2001.

3   Kaitin, K.I. (ed.) Cost to develop new biotech products is estimated to average $1.2 billion. *Tufts Center for the Study of Drug Development Impact Report*, 2006. Vol. 8.

ing the time and cost of development.

Once a drug receives regulatory clearance for marketing, it will likely be protected by patents that were filed before clinical trials began. With an average of twelve years of patent protection remaining after FDA approval, marketing and sales efforts must generate revenues and expand market penetration to deliver a return on R&D expenditures. After a drug is on the market, drug sponsors must monitor patients for unexpected side effects. Independent or sponsored clinical trials can also test suitability for additional indications, potentially expanding the market. The emergence of competing products and looming patent expiration dates motivate the development of alternative drug forms and formulations to leverage established brands, and modification of marketing methods to extend sales. Some firms specialize in modifying patented drugs, capitalizing on the negative specter of patent expirations, and patent their modifications to license them to pioneers as a means to preempt generics.

Box

The Human Genome Project and drug development

The Human Genome Project (HGP) was a multi-billion dollar multinational effort to sequence the entirety of the human genome, identify all the genes, improve tools for genome analysis, and address related ethical, legal, and social issues. The project started in 1990 and sequencing was completed ahead of schedule in 2003.

The initial findings from the HGP informed scientists that molecular biology was far more complicated than previously believed. Projections for the number of genes in the genome, for example, ranged from the high tens-of-thousands to more than 100,000. After examining the sequence of the human genome, it was found that the genome contained far fewer genes than previously suspected—less than 30,000. The key to enabling the genome to produce a sufficient diversity of proteins from this relatively small set of genes is in editing individual gene mRNAs so that each gene can produce multiple proteins. Entirely new methods for regulation of gene expression were also discovered, further complicating efforts to tame molecular biology.

So, why has the HGP not yet produced a revolution in drug development?

- The HGP was primarily a basic science endeavor. It has greatly expanded the knowledge-base essential for drug development
- Information gleaned from the HGP must be interpreted and understood before it can produce new drug leads
- It can take more than a decade for a drug lead to gain regulatory approval
- Most drug leads fail to gain regulatory approval

While processing the new information from the HGP will occupy researchers for decades to come, there are some immediate benefits. The project brought many technological advances. The costs of synthesizing and sequencing DNA, for example, decreased by several orders of magnitude and the precision of many experiments has improved, along with the ability to automate many procedures. These improvements have translated beyond humans into agriculture, where farmers are better equipped to identify top-performing plants and select them for traditional breeding programs. DNA fingerprinting, which has greatly advanced the field of forensics, is also a spin-off of the human genome project. These advances and others are benefitting science today, while we await the other outcomes of the HGP.

Time and cost of whole-genome sequencing

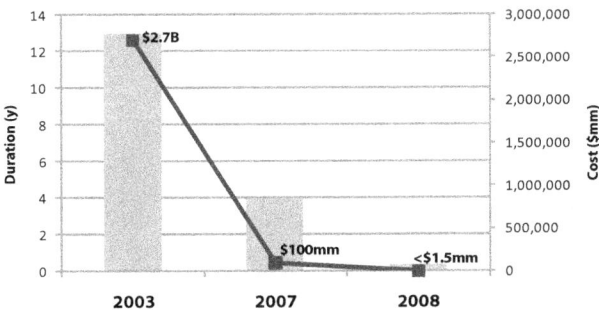

Figure 4.6 *Declining time and cost of human genome sequencing*
Source: James Watson's genome sequenced at high speed. *Nature*, 2008. 452:788.

The aforementioned five steps of drug development must always occur, and the influences of individual biotechnology innovations are compartmentalized. Each innovation can only affect one or a few steps. For example, functional genomics can aid in developing model systems and selecting or designing potential drugs, but cannot resolve drug delivery or manufacturing issues. Molecular evolution can refine lead compounds, but cannot assist discovery, clinical trials, or manufacturing. The impact of this compartmentalization is that no single technology can profoundly alter the process of drug development; a number of complementary innovations are necessary for a revolution.

QUALITIES OF A "GOOD" DRUG

Beyond safety and efficacy, many other factors influence the quality of a drug. An ideal drug must address a market that is willing to pay a price that permits profitable sales. The commercial attractiveness of a drug is influenced by the size of a drug's patient population, the frequency of dosage, cost of production, barriers to entry of competitors, and availability and cost of alternative treatments.

Consider the example of penicillin, the first antibiotic. The devastating effects of bacterial infections and the absence of effective treatments guaranteed strong sales. The raw materials for penicillin production were known and initial tests showed a good safety profile. The remaining challenge for penicillin production was developing a method to produce sufficient quantities of the drug at a reasonable price.

In addition to safety and efficacy considerations, practical aspects such as chemical stability and therapeutic administration also affect the commercial prospects of a drug. Oral administration and patches are generally preferred to injections. Unfavorable administration can reduce patient compliance or place a drug at a competitive disadvantage to alternatives. Some drugs must also be mixed with carriers, which can impart their own side effects (see *Drug Delivery* in Chapter 5).

The chemical stability of a drug affects the conditions under which it can be distributed to pharmacists and stored by patients (e.g., Is it heat sensitive? Must it be refrigerated?). The

Table 4.1 *Qualities of a "good" drug*

Market quality	Pharmacological qualities	Barriers to entry
• Market size • Dosage frequency • Manufacturing cost • Price elasticity	• Safety • Effectiveness • Chemical stability • Metabolic stability • Drug delivery	• Intellectual property protection • Market exclusivity • Challenge of generic manufacture

metabolic stability determines how long the product remains effective in a patient's body and how much product must be administered to achieve a therapeutic effect. Metabolic stability also influences how many daily doses are necessary, and it can also be a factor in side effects. A drug's solubility affects numerous outcomes: if it will be stored in fat, whether or not it can travel through the bloodstream, if it can cross the gastrointestinal lining or be delivered through the skin, and if the kidneys can excrete it. These factors impact drug delivery, dosage regimens, and the potential for side effects. Toxicity or side effect limits may be relaxed if the alternatives to drug treatment warrant it.

Copying drugs is a sufficiently lucrative business to motivate companies to specialize in challenging patents and developing generic versions of drugs with expired patents. To prevent competitors from capitalizing on the efforts of pioneers and selling drugs for prices which reflect their reduced R&D burden, drugs must have some form of commercial protection. Drugs may be protected by patents, trade secrets, or other methods such as FDA-granted temporary market exclusivity. The vehicles that offer exclusive market rights for drugs are designed to promote innovation by granting drug developers temporary monopolies that permit them to recoup their investments in research and development.

Chapter 5

Tools and Techniques

I think the evidence is overwhelming that one tool doesn't do it. Yet time and time again, we see new entrants coming into this business saying, "This tool will revolutionize the discovery process," when it's much more likely that the integration of tools, of how they work, will have a much more powerful effect.
Harvard Business School Professor Gary P. Pisano

I t is only through an understanding of the tools and techniques used for biotechnology research that one can develop an appreciation of the possibilities and challenges of biotechnology. This chapter presents a survey of biotechnology tools and techniques to foster such an appreciation.

The tools and techniques used for biotechnology research define the universe of products and services that biotechnology companies can develop. While a biotechnology company could decide to focus on a set of applications or a set of technologies, most define themselves around technologies rather than applications. From a practical perspective, it is simpler to focus on a technique that can have several applications than to search for all the methods to solve a single problem. Whereas tackling a specific application may require expertise in numerous techniques, a single technique can be used for multiple applications, positioning research and development to serve multiple markets reduces market risks associated with any individual market. Therefore, it is simpler and often preferable to develop expertise and patents for a few techniques and exploit this competitive advantage to develop solutions for multiple applications.

BIOINFORMATICS

Bioinformatics is the convergence of information technology and biotechnology, applying information technology to manage and analyze the vast amounts of data generated from basic biological research. Bioinformatics assists scientists in managing data and enables interpretation of data by presenting it in useful formats.

The tools and techniques that define bioinformatics are themselves a demonstration of the growth and diversity of techniques in drug discovery. As late as the mid-20th century, drug discovery was conducted mainly through chemical synthesis followed by extensive trial-and-error testing. Large-scale testing of derivatives of potential drugs was introduced in the 1970s, followed by attempts at rational drug design in the 1980s. Bioinformatics entered the arena in the 1990s, enabling drug synthesis and testing to be simulated by computers. As biological knowledge, computational power, and computer algorithms improve, it becomes increasingly possible to identify and refine potential leads through the use of computers.

There are two successive elements in bioinformatics: data assembly and data analysis. Computer-assisted data management enables gathering, analysis, and representation of biological information, helping scientists better understand biological processes, understand the mechanisms behind diseases, develop methods to treat diseases, and develop other applications based on biological knowledge. Bioinformatics also allows researchers to perform comparative and predictive studies of biological processes. Applications of bioinformatics data analysis include prediction of protein structure, prediction of protein function, and drug target selection.

One important development that emerged at the beginning of the twenty-first century was the implementation of automated research techniques such as DNA sequencing and robotic fluid-handling and assay systems, permitting large-scale research efforts. A defining event in the automation of biological research was sequencing the human genome. This mammoth project, extending over a decade, determined the sequence of the three billion base-pairs that comprise human

DNA. The logistical problems of collecting and managing this mass of information required the development and application of novel computer technologies to assist biological research.

Sequencing the three billion base-pairs of the human genome was only possible with the use of bioinformatics to manage all the data. The information that can be processed using bioinformatics techniques includes not only sequence information for genes and proteins, but also details on the structure and function of proteins, disease correlations, and raw information produced from scientific experiments such as microarray analyses and protein interaction studies.

Bioinformatics applications exist for most steps of drug development. Predictive and analytical algorithms can screen potential lead compounds or help design them from scratch; toxicity can be predicted by comparison against compounds with known properties; even clinical trials can be simulated. Any biological information that can be entered into a computer database is subject to processing and analysis by bioinformatics.

A strength of bioinformatics is the ability to extract information and identify patterns from large databases. Data mining, an analytical bioinformatics application, uses computers to analyze masses of information. Integration and comparison of numerous experimental observations permits the discovery of trends and patterns in large databases, potentially identifying novel relationships.

CLINICAL MODELING

Clinical trials are necessary to demonstrate that drugs are safe and effective. Drug developers are not permitted to market drugs until they receive FDA approval, based on safety and efficacy data generated in clinical trials. Time spent in clinical trials is time that cannot be spent selling drugs. Because drug patents must be filed prior to the initiation of clinical trials, any reduction in clinical trial duration can be valuable—a single day's delay in approving a billion dollar blockbuster drug can mean a loss of revenue exceeding $3 million.

The processes for clinical trials are flexible—they can be adapted for diverse drugs—positioning clinical research organizations to build businesses around this critical step in drug development. By focusing on clinical trials, contract research organizations are able to develop specialized expertise and relationships with clinical trial providers. Clinical research organizations generally offer two distinct solutions to aid clinical trials. Trial management solutions involve strategies to design trials, facilitate patient recruitment, speed and improve communication between trial investigators and sponsors, and manage and analyze data. Simulation and prediction services use sophisticated computer techniques to enable the safety and efficacy of drugs to be predicted at a lower cost and with greater speed than actual clinical trials, ultimately permitting selection of compounds and trial protocols most likely to lead to FDA approval.

FUNCTIONAL GENOMICS

Functional genomics seeks to understand the activities of genes in healthy and diseased states. The reality of human genetic variation means that different patients respond differently to the same drug. The difference may be as simple as a slight difference in efficacy or it may result in a drug being completely ineffective or even toxic in some patients. It is estimated that most commonly used drugs are effective in only 30–60 percent of patients with a given disease. A subset of these patients may suffer severe side effects.

Without functional genomics there is no simple way to determine if a given patient or subset of the population is likely to respond either well or poorly to a medication. As a result, drugs are developed for the "average patient." Furthermore, many drugs that might benefit a subset of patients may never be developed because they cannot be shown to be useful in an average group of patients. Patients who are unlikely to benefit from a drug or who may suffer adverse side effects are likewise not identified and given more appropriate treatments. Functional genomics enables segmentation of patient groups to resolve these issues.

Knowledge of the sequence of the human genome is a valuable tool for functional genomics. However, simply knowing the sequences of genes is not sufficient. Discrete genetic differences between individuals must be correlated with the effects of medications. Studying single nucleotide polymorphisms (SNps) and pharmacogenetics reveals correlations that enable functional genomics.

SNps are discrete DNA sequence changes between individuals that are at the root of many genetic differences. SNps have been linked to the likelihood that an individual will find a drug effective or unsafe. These therapeutic differences are related to variations in drug targets, in enzymes that metabolize drugs, and in other molecules involved in cellular metabolism. The elucidation of discrete genetic differences that can be readily identified holds the potential to predetermine how a patient will respond to a drug.

PHARMACOGENETICS & PHARMACOGENOMICS

Pharmacogenetics studies the relation between genetic variation and the effects of pharmaceuticals: the investigation of how genetic differences affect the ways in which people respond to drugs. Specifically, pharmacogenetics seeks to understand the differences between drug targets and metabolic enzymes that affect efficacy and toxicity. Differences in genetic sequences are responsible for many of the differences between individuals. Just as genes influence eye color and hair color, they can also influence susceptibility to disease and determine whether specific drugs are safe and effective for certain individuals. The terms pharmacogenetics and pharmacogenomics are sometimes used to respectively distinguish between the correlation of single drugs with multiple genomes, and of multiple drugs with single genomes.

Learning why certain individuals are unresponsive to drugs or experience dangerous side effects gives researchers the potential to develop drugs that address these shortcomings. Studying the mechanisms by which drugs are rendered ineffective or toxic may also enable drugs to be designed to avoid or compensate for these alterations. Furthermore, drug discovery cost and time can be reduced by eliminating potential clinical trial participants for whom drugs in development are likely to prove ineffectual. More precise clinical trials justify smaller

Box

Cytochrome p450 and pharmacogenomics

Cytochrome p450 is a generic term for a set of enzymes which are collectively the most important element in chemical modification and degradation of chemicals including drugs and other foreign compounds. A vast majority of the most serious adverse reactions to medicines appear to involve drugs that are metabolized by the cytochrome p450 system.[1]

Six different p450 genes are responsible for most of the metabolism of commonly used drugs. Each gene can have dozens of discrete mutations affecting its activity. Inventorying the set of cytochrome p450 enzymes and elucidating the factors contributing to their expression and activity levels is central to understanding and predicting differences in response to drugs. Some of the drugs and compounds degraded by cytochrome p450 enzymes are caffeine, morphine, Taxol, Prilosec, cocaine, codeine, Viagra, St. John's wort, and HIV protease inhibitors. If two compounds are degraded by the same cytochrome p450 enzyme it is possible that taking both compounds at the same time can lead one or both to accumulate to dangerous levels. This is one of the ways in which drugs can interact to alter efficacy or have lethal consequences.

Roche's AmpliChip 450 is the first microarray-based diagnostic test that can detect genetic variations influencing drug efficacy and adverse drug reactions. The AmpliChip contains 15,000 DNA sequences representing 31 genetic variations in two cytochrome p450 enzymes. According to Roche, the two enzymes affect 25 percent of commonly prescribed medications. The purpose of the chip is to determine whether a patient metabolizes drugs at a normal, slow, or fast rate. This information can help doctors prescribe appropriate medications and dosages based on a patient's rate of degrading specific drugs. Properly calibrated dosages can mean the difference between no response, therapeutic effectiveness, and serious side effects.

---

1   For a topical review, see: David A. Katz, D.A., Murray, B., Bhathena, A., Sahelijo, L. Defining drug disposition determinants: a pharmacogenetic–pharmacokinetic strategy. *Nature Reviews Drug Discovery*, 2008. 7:293-305.

and fewer trials, facilitating FDA approval (see commentary on Herceptin clinical trials in Box *Personalized medicine and drug sales* in Chapter 6).

Functional genomics can facilitate drug discovery and improve drug administration. Applying functional genomics in a personalized approach to medicine—prescribing drugs only to patients likely to benefit from them—can streamline medical care and avoid unnecessary side effects. Functional genomics can also potentially identify patient groups of less than 200,000 Americans, qualifying drugs for Orphan Drug status, which grants special incentives and market exclusivity.

MICROARRAYS

Microarrays are tools that enable the identification of DNA or other samples and examination of gene expression and protein modifications in individual tissues, and under different conditions. While the first microarrays were directed at detecting DNA, new technologies have enabled protein-based and other forms of microarrays. Microarrays enable researchers to detect the presence or expression of many genes and proteins at once. The ability to simultaneously examine the changes in expression of many different genes, or changes in protein levels and modifications, is useful in investigating the effects of diseases, environmental factors, drugs, and other treatments in human health.

Applications of microarrays include diagnosing or identifying cancerous cells, assessing genetic predispositions to diseases, examination of gene expression, and gene and protein responses to drugs or other therapeutic procedures.

PROTEOMICS

Proteomics is the study of protein structure and function. Genes encode proteins, which perform structural and functional roles in cells. It is proteins, not genes, which are the major actors in molecular biology (see Chapter 3). Understanding the structure and function of proteins can lead to new therapies and influence disease diagnosis and treatment.

Proteins can be roughly categorized by their structural and functional roles. An example of a structural protein is keratin, a component of skin. Functional proteins, called enzymes, perform cellular duties. Metabolic enzymes aid in food digestion and enable harvesting of stored energy. Studying DNA can reveal some information on the control of protein synthesis, but provides limited information about the structure and function of proteins. This requires examination of the proteins themselves.

Proteomics uses a variety of techniques to examine protein structures and functions. Unlike DNA sequencing, which is a relatively uniform technique that can be widely used without modification, the very methods used to investigate proteins vary with each individual protein being studied. There is no simple or uniform way to produce, identify, quantify, or characterize proteins. The need to continually adjust experimental methods in proteomics research is a significant challenge to scaling or automating research efforts.

MANUFACTURING

Developing a drug that is safe and effective is essential to gain FDA approval, but to generate revenues and recoup development costs it is necessary to manufacture and sell the product as well. Conventional wisdom once held that investing in manufacturing process development did not benefit a company's returns as much as basic research. In an era of increased competition where companies frequently produce competing treatments, the ability to accurately predict demand, to rapidly develop production methods, and to scale production capacity provide a strategic advantage.

In the process of drug development a drug may be produced in test tubes, flasks, small fermentation vessels, pilot-plants, and large-scale production facilities. The progression from bench-top production to large-scale production is not a trivial process. As the scale of production increases, factors such as temperature, oxidation, and mixing change, potentially altering the final product. To ensure drug quality, manufacturers must demonstrate compliance with FDA current good manufacturing practices (cGMP) and further prove that drugs of consistent purity and activity can be produced in large quantities from batch to batch, day after day, year after year.

One alternative to traditional manufacturing methods is the use of animals and plants that are genetically modified to produce a desired compound. For example, drugs may be harvested from chicken eggs or cow milk, or purified from plant tissues. A significant benefit of transgenic production is that production of raw material is relatively simple to implement and scale. It is estimated that plant-based biologic production can be 10 to 1000 times less expensive than conventional fermentation systems. Leveraging the relative simplicity and cost advantage, drug-producing varieties of animals and plants can be distributed in regions lacking sufficient expertise, facilities, or resources for conventional fermentation production. To scale production, these transgenic factories can be bred or cloned, increasing the supply of raw product.

DRUG DELIVERY

Despite the emphasis on the biological activity of drugs, it is important to also consider the systems and methods used to deliver drugs to their therapeutic targets. The goal of drug delivery systems is to enable active medications to reach appropriate parts of the body, in the appropriate concentrations for the appropriate amount of time, where they can accomplish their therapeutic task.

Drugs produced by biotechnology techniques tend to be large proteins and nucleic acids which face special challenges relative to smaller, more chemically stable pharmaceutical drugs. Factors impeding oral delivery of biologic drugs include:

- Acidity of the digestive system
- Intestinal enzymes that degrade proteins
- Inability of biologic drugs to cross intestinal walls
- Poor solubility of biologic drugs

Overcoming the challenges of biologic delivery can present new opportunities, as targeted and metered dosage systems can potentially improve drug effectiveness, mitigate safety and side effect concerns, and ultimately improve patient compliance and retention.

Patient compliance is also a concern in drug delivery. The requirement to take many pills a day, to follow rigorous dosage regimens, or the use of unappealing delivery methods such as injection may discourage compliance, and patients cannot benefit from drugs they don't take. Reducing administration from several times a day to once a week by using an extended-release formulation, for example, can dramatically improve compliance and ultimately improve patient outcomes—the primary objective of drug therapy.

Delivery techniques that increase compliance can ultimately help a company derive more revenue from a product. Selling twice as much of a drug by doubling the duration that patients take the drug or doubling the number of people who take it is arguably similar to selling two drugs, without the cost of developing two drugs.

NANOTECHNOLOGY

Nanotechnology is a multidisciplinary field encompassing the development and application of materials at sizes measured in billionths of a meter. Surface tension, molecular interactions, and surface area exposure play an increasingly important role in chemical and physical interactions at this size range, giving nano-scale materials properties that are markedly different than those seen at larger scales. Nano-sized flour particles, for example, are capable of igniting violently and causing explosions in flour mills. Geckos are able to climb walls because of nano-scale hairs on their feet which use atomic interactions, rather than stickiness, to adhere. The enzymes and other key molecular players in biotechnology also operate at this size scale, creating an opportunity for convergence between biotechnology and nanotechnology.

As with biotechnology, early investments in nanotechnology research were attracted by the high profit potentials of serving unmet medical needs. Beyond developing new therapeutic products with nanotechnology, there are also strong

Table 5.1 *Selected nanotechnology applications in drug delivery*

Technology	Benefit
Carriers	Improve solubility and avoid need for harsh solvents
Encapsulation	Extend duration of drug bioavailability
Nanoparticles	Improve solubility, speed delivery, and extend duration of drug bioavailability

opportunities in developing delivery systems that can improve the safety and efficacy of existing drugs (see *Drug Delivery* earlier in this chapter). Reducing the particle size of drugs has the potential to:

- Increase surface area
- Enhance solubility
- Improve oral bioavailability
- Speed onset of therapeutic effect
- Decrease necessary dosage
- Decrease variability between fed and fasted dosage
- Decrease patient-to-patient variability

Abraxane is an example of how nanotechnology can improve existing drugs. It was developed as an improved version of Taxol. The castor oil carrier required to solubilize Taxol (see *Drug Delivery* earlier in this chapter) has several significant side effects associated with it. By binding paclitaxel (a generic version of Taxol) to nano-scale protein particles, Abraxis Oncology produced a version of the drug that could be injected without castor oil, dramatically improving the side effect profile.

Nanotechnology also has many applications in research. Researchers continually seek to perform assays using smaller and smaller quantities of experimental materials, which are often very expensive and difficult to obtain. Smaller-scale experiments provide two advantages: the ability to perform more experiments at lower costs, and the ability to run more experiments in less space. Microarrays, for example, enable researchers to examine many genes at once using a small quantity of material. Nanotechnology-based solutions, such as "lab on a chip" products, can reduce the scale of experiments even further, enable multiple sequential or simultaneous assays, and expand opportunities for automation in research.

Chapter 6

Applications

The best way to predict the future is to invent it.
Richard Feynman

Biotechnology companies focus on selling products or offering services. Products include drugs, reagents, research tools, industrial enzymes, and specialized crop plants. Services include discovering drug lead compounds, clinical trial management, and manufacturing.

The most common application for biotechnology companies is drug development. This is partially due to the enormous profit potential of drugs, which can greatly offset the increased development cost relative to other applications. Some firms specialize in elements of the drug development process, supporting the search for potential drug compounds by coordinating clinical trials or producing research tools and drug delivery systems. These "pick and shovel" companies benefit indirectly from the profits of other development companies by selling necessary products and services. While some service firms fulfill functions that could be developed internally, dedicated service firms may also possess economies of scale, enabling them to offer specialized expertise and abilities.

Drug development differs from most other commercial development ventures because products must be proven safe and effective before they can be marketed. Applications not intended for human use benefit from not requiring clinical trials to prove their safety and efficacy prior to commercial release, although these applications may still be subject to EPA, USDA, and other political or ethical restrictions.

Table 6.1 shows the division of biotechnology applications into three categories: Green biotechnology for agricultural ap-

Table 6.1 *Biotechnology application categories*

Category	Description
Green: Agricultural biotechnology	Products and applications related to livestock and crop production, and agricultural production of biotechnology products.
White: Industrial biotechnology	Modification or improvement of industrial processes such as paper processing, bioremediation, and chemical and organic compound synthesis.
Red: Medical biotechnology	Drugs and other agents to treat, cure, or prevent disease, and products that assist in the diagnosis of diseases or measurement of crucial factors in health and disease.

plications, white biotechnology for industrial applications, and red biotechnology for therapeutic applications. Specific applications are described in further detail in this chapter.

In reading this chapter, it is important to recognize that biotechnology is not a panacea. In addition to practical considerations such as technological constraints, the legal, regulatory, political, and commercial factors described later in this book have profound impacts on the ability to develop and commercialize biotechnology. Many biotechnology companies fail because they develop products for which profit-enabling markets do not exist.

GREEN BIOTECHNOLOGY: AGRICULTURE

The directed modification of plants and animals can increase their value in agricultural applications. By studying the genes responsible for specific traits, it becomes possible to introduce, alter, or change the expression of those genes in a controlled manner, resulting in a desired change. Extensive testing, mandated by the FDA, EPA, and USDA, is required to determine if genetically modified plants are safe for humans and ensure that they do not pose a threat to the environment. This testing requires demonstration that foreign proteins in edible crops are decomposed by cooking or stomach acids, precluding their ability to cause adverse effects if ingested.

Traditional agriculture relies on crossbreeding and hybridization to improve the quality and yield of crops and domesticated animals, and to overcome natural obstacles such as disease. These methods involve controlled breeding of plants and animals with desirable traits to produce offspring that ideally will retain the best traits of the parent organisms. Virtually every plant and animal grown commercially for food or other uses is a product of crossbreeding and/or hybridization. Relative to biotechnology methods, these processes are costly, time consuming, inefficient, and subject to significant practical limitations.

Genetic modification of crops has produced herbicide resistant strains, insect resistant strains, enriched foods, and improved industrial products. Herbicide and pest resistant crops can have a profound positive impact on the environment by making it possible to raise crops using dramatically less pesticide. Transgenic corn containing insecticidal toxins from *Bacillus thuringiensis* (Bt) bacteria can prevent corn borer infestations without chemical crop dusting that is toxic to humans and also kills beneficial insects. Producing these herbicide and pest resistant crops by traditional methods—if it were possible at all—would take dozens of generations.

A study of the first ten years of commercial genetically modified crop growth—from 1996 to 2006—found economic benefits to farms of $5 billion in 2005 and $27 billion over the ten year period. Pesticide use was also reduced by 224 million kilograms, or 6.9 percent, resulting in a reduction of environmental impact by more than 15 percent.[1] While improved seeds may cost significantly more than conventional seeds, it is estimated that conventional farmers spend significantly more on chemical insecticides and herbicide than they spend on seeds. Furthermore, genetically modified seeds have helped improve yields significantly; U.S. corn harvest yields have doubled since 1970, and Monsanto predicts that harvest yields will double again by 2030 (see Figure 6.1).[2]

The most abundant genetically modified crops are cotton,

1 Brookes, G. and Barfoot, P. Global Impact of Biotech Crops: Socio-Economic and Environmental Effects in the First Ten Years of Commercial Use. *AgBioForum,* 2006. 9(3):139-151.

2 Hindo, B. Monsanto: Winning the Ground War. *BusinessWeek*, December 17, 2007.

Corn harvest yields

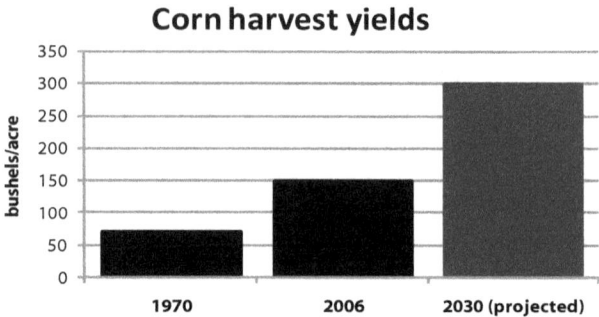

Figure 6.1 *Progress in agricultural yields*
Source: Monsanto

corn, soy, and canola. More than 1 billion acres of genetically modified crops had been sown in 17 countries by the end of 2005, a decade after Monsanto introduced the first genetically modified crop. According to the USDA, 52 percent of all corn, 79 percent of upland cotton, and 87 percent of soybeans planted in the United States in 2004-05 were biotechnology-derived varieties. The International Service for the Acquisition of Agri-Biotech Applications reports that in 2007 the number of farmers growing biotechnology-derived crops exceeded 12 million—11 million of whom were defined as resource-poor farmers—and hectarage exceeded 114 million acres.

CHALLENGES

Techniques to insert genes into plants are well established, but a remaining challenge for agricultural biotechnology is the realization of desired modifications. With many plants sporting genomes larger than humans, a fundamental understanding of agricultural biology is necessary for application. Confounding efforts at genetic modification, introduced genes may be unstable, and unforeseen biological issues may interfere with introduced proteins.

Another challenge for agricultural biotechnology is public acceptance. Because public support is essential to enable application of biotechnology, it is important to be sensitive to potential objections and to encourage positive perceptions. Public concerns can impact political decisions, resulting in bans on crop plantings or even sales of crops.

Critics of genetically modified foods warn that inserting genes into plants and animals may have unforeseen results, increasing the risks of allergic reactions to foods and other health problems. The potential for genetically modified plants to crossbreed with wild stocks or to cause environmental damage is also a prevalent concern. It is worth noting that many of these same concerns would also preclude the use of traditional farming practices and breeding methods if they were applied beyond biotechnology.

Beyond answering critics of genetic modification, developers must also find markets with a preference for their products and who are also willing and able to pay a profitable price. A case example of failure to find a profit-enabling market is the Flavr Savr tomato, a tomato engineered to have a longer shelf-life. In 1994 the FDA approved the Flavr Savr tomato, the first genetically modified whole food product. Interestingly Flavr Savr tomatoes were pulled from the market not because of consumer resistance, but rather due to customer disinterest and an inability to sell for a profit.

While improved yields, nutritional enhancement, and the ability to grow crops on marginal soils may appear to individuals in developed nations to be trivial or cosmetic improvements, but the situation in developing countries makes these improvements far more imperative. As population encroachment and environmental change are decreasing the quantities of arable land in many parts of the world, aging distribution infrastructures are challenged to deliver food to ever-increasing populations. For many developing nations the ability of biotechnology to increase crop yield, reduce farming inputs, and expand arable land presents a vital solution to the need to grow more crops on less land in order to prevent otherwise-inevitable widespread starvation.

TREE BIOTECHNOLOGY

An early goal of forest scientists has been to produce trees with less lignin, avoiding expensive and environmentally toxic procedures in paper production. Removal of lignin in paper production requires an enormous amount of energy and chemicals, making the pulp and paper industry the second most energy-intensive industry group in the U.S. manufacturing sector. Another problem is that forestry demand for wood products exceeds the rate of renewal. Some tree species require well over a century to reach economic viability. Using genetic engineering to develop faster, straighter, and taller-growing trees can potentially fill a market need while preserving old growth forests.

ANIMAL BREEDING AND CLONING

Biotechnology can serve two roles in animal breeding. First, it can enable the use of genetic markers to identify desired animals for breeding programs. Biomarkers associated with characteristics such as milk production, meat quality, or hereditary diseases can be used to inform traditional breeding programs and improve herd quality. The second, more contentious application, is to directly clone desired animals.

In 2008 the FDA released a "final risk assessment," concluding that foods from healthy cloned animals are as safe as those from traditionally-bred animals. Cloning animals costs substantially more than traditional or assisted breeding programs, but this additional cost is somewhat offset by the near-certainty that the clone will inherit desired traits. The extra cost of producing cloned animals also provides an assurance to wary consumers—the high cost of producing these animals means that they are unlikely to be used as meat or milk sources, but will likely be relegated to breeding.

FUNCTIONAL FOODS

Functional foods contain elements which can provide health benefits beyond simple nutrition. Existing examples include high omega-3 eggs, produced by feeding ground flax seeds to hens, and supplementing processed foods with soy protein or fiber.

Beyond improving yields, biotechnology also has the potential to dramatically impact the nutritional qualities of food. Nutritional modifications to plants include conferring the ability to synthesize essential vitamins, reducing the undesirable saturated fat content of cooking oils, increasing protein quan-

---

Box
Carnivorous fish as vegetarians

Fish farming, or aquaculture, has been heralded as a potential solution to overfishing of wild stocks. The situation is complicated, however, by the need to satisfy carnivorous fish diets. Carnivorous fish aquaculture accounts for the majority of global fish oil usage, and is rapidly growing to become the primary market for fish meal as well—aquaculture is depleting wild fish stocks.

Enter biotechnology. By studying the protein requirements of carnivorous fish, researchers can potentially use traditional breeding and genetic modification to produce terrestrial crops as feed. These crops can both address the nutritional needs of fish, and can resolve some of the environmental issues related to aquaculture. Furthermore, growing plant crops to feed fish farms creates new revenue opportunities for farmers and holds the potential to enable inland farmers to farm fish, improving the distribution and availability of fresh fish.

Despite its promise, it remains to be seen whether traditional breeding can produce plants that can satisfy the nutritious needs of fish, if consumers will accept fish products that have been fed genetically modified plants, if the quality and flavor of plant-fed fish will match fish-fed or wild fish, and if plant-fed aquaculture can be economically feasible.

---

tity and quality in vegetable staples, and reducing allergenic properties of milk, wheat, and other products.

These nutritional improvements can have a significant effect on human health. For example, enabling staple crops such as rice to synthesize vitamin A precursors or to make iron bioavailable respectively hold the potential to prevent blindness and anemia in countries where commercial vitamin supplements are unaffordable. Calgene's (now owned by Monsanto) Laurical is the first commercially available functional food oil, approved by the FDA in 1995. While conventional canola oil does not contain lauric acid, Laurical contains 38 percent lauric acid. Laurical has applications in soaps and detergents, chocolate, low-fat coffee whiteners, and imitation cheeses.

MOLECULAR FARMING

Molecular farming produces useful products from domesticated plants and animals through genetic engineering. Pharming is a subset of molecular farming and produces therapeutic drugs using genetically altered animals and plants. The distinction between molecular farming and traditional farming is that the plant and animal products of molecular farming are not eaten as food, but are harvested to produce useful biotechnology products. Safeguards to prevent exposure through accidental ingestion include sequestration of crops (plans have included growing plants in abandoned mine shafts), production of recombinant proteins in non-edible portions of plants, and expressing proteins at very low levels, requiring extensive processing to obtain measurable and useful quantities of recombinant materials.

A non-therapeutic application of molecular farming is the mass-production of spider silk. Stronger than steel and lighter than cotton, spider silk manufacturing has traditionally been impeded by the inability to domesticate spiders. Nexia Biotechnologies has produced dragline spider silk in laboratory conditions. Their intention is to spin spider silk produced in the mammary glands of genetically modified goats for use in fishing line and in military applications such as lightweight body armor.

PHARMING

Many developing countries that could benefit from commercially available therapies for diseases such as hepatitis and other endemic conditions cannot afford to purchase appropriate medicines or produce them locally. Production of therapeutic vaccines in familiar crops such as bananas and potatoes, or chicken eggs, can enable local farmers to manufacture medicines without the need for sophisticated production techniques or expensive purification methods.

The traditional method for manufacturing biological drugs is fermentation in huge stainless steel vats. A significant advantage of pharming is that it can decrease the cost of drug manufacturing. Furthermore, whereas scaling fermentation systems requires building and receiving FDA approval for additional facilities, boosting pharmed drug production may be as simple as planting more transgenic plants or increasing the size of a transgenic animal herd. This is especially important in countries where expertise and facilities for large-scale fermentation are not available. These advantages are offset by the higher up-front costs and longer lead-times required to produce transgenic animal or plant production systems (see *Manufacturing* in Chapter 5).

The development of drugs that are easy to purify or that can be administered without purification is essential for enabling pharming. Application of pharming is challenged by the threat of uncontrolled spread of genetically engineered plants and animals, and by start-up development costs. In the case of drug manufacturing, it is preferable to retain formulation flexibility and delay the bulk of manufacturing expenditures until a drug's safety and efficacy have been assessed. The financial cost and time required to develop transgenic production systems are at odds with this strategy.

WHITE BIOTECHNOLOGY: INDUSTRIAL PROCESSES & BIO-BASED PRODUCTS

Industrial biotechnology is the application of molecular biology techniques to improve efficiency and reduce the environmental impacts of industrial processes. Just as biotechnology has transformed agriculture, drug discovery, and development, it can similarly affect industrial operations.

Industrial biotechnology companies develop biocatalysts such as enzymes that are used for chemical synthesis. Enzymes are a category of proteins which are produced by all living organisms (see *Proteins and Enzymes* in Chapter 3). Enzymes enable the biochemical reactions necessary for life by increasing reaction rates. In biological systems, enzymes help digest food, assemble complex molecules, and perform other complex functions. Specialized enzymes are also used extensively as detergents as well as in the production of beer, cheese, and fruit juice. Bacteria have developed specialized enzymes that allow them to live in a wide variety of extreme environments; from thermal vents at the bottoms of oceans to the insides of rocks. Enzymes are characterized according to the compounds they act upon. Some of the most common enzymes with industrial applications are proteases, which break down protein; cellulases, which break down cellulose; lipases, which act on fatty acids and oils; and amylases, which break starch down into simple sugars.

By studying diverse bacteria and other organisms, scientists discover novel biocatalysts that function optimally under a wide variety of conditions, including the relatively extreme levels of acidity, salinity, temperature, or pressure found in some industrial manufacturing processes. In other cases, enzymes can remove the need for extreme conditions or harsh chemicals, saving energy and reducing environmental impact.

The application of biotechnology to industrial processes is appealing because of the potential to affect yield, effectiveness, and production cost of products with established markets. Serving established markets cam improve the accuracy of market size projections, helping justify high R&D investments and attracting funding for large market opportunities. The potential for application of biotechnology to reduce infrastructure requirements may also make it possible to profitably address smaller markets. For example, see the example of *Oil Well Completion* below. An additional appeal for industrial

---

Box

Blue jeans and biotechnology

1.8 billion pairs of denim jeans are sold each year, making them among the most prevalent clothing items sold worldwide. *Stonewashing* is commonly used to soften the jeans and fade the dyes to give the jeans a slightly worn appearance. This process was traditionally performed by tumbling jeans in large machines with abrasive pumice stones. This process can weaken jeans and damage washing machines, and requires several rinsings to remove all the pumice traces.

An alternative enzyme-based method has been introduced which imparts several benefits. The degree of stonewashing can be attenuated by using cellulase enzymes to break down the denim cellulose fibers in a controlled manner. This process also requires less water and energy than traditional stonewashing, and results in longer-lasting jeans.

biotechnology development is the greatly reduced regulatory burden relative to pharmaceutical applications.

BIOFUEL AND LUBRICANTS

Petroleum prices, political considerations, and the threat of shortages all motivate the search for alternative fuel sources. Processes to convert cornstarch into ethanol, a petroleum additive and alternative, have been available for many years, but questions about the scalability and ultimate economics of this approach have led to the search for alternative methods to produce ethanol.

Fuels derived from petroleum are the product of compression and heating of prehistoric plants and animals over geological time scales deep below the earth's surface. Because of their ultimate biological source, the potential exists to use alternative processes to make fuels from sources such as plant materials and animal fats in less time. There are three basic methods to produce fuels from plant and animal sources: chemical transesterification, fermentation, and cellulose degradation. These fuels can be used in cars and trucks, as well as in numerous other applications.

Transesterification is a process used to convert vegetable oils, animal fats, and recycled greases into biodiesel. This is technically a chemical process, not an application of biotechnology. Fermentation is the use of bacteria or yeast to convert simple sugars, abundant in plants such as corn and sugar cane, into ethanol. This is fundamentally the same process that is used to make beer, wine, and other alcohols. Cellulose degradation significantly expands the prospects for fermentation by enabling the use of a wide variety of feedstocks. Whereas fermentation requires feedstocks with a high content of simple sugars, cellulose degradation uses chemical pre-treatments and cellulase enzymes to break down cellulose, a complex sugar, into simple sugars. These simple sugars can then be used in traditional fermentation to make desired chemicals.

The principal advantage of cellulose degradation over other methods is the abundance of cellulose in materials such as farm waste, wood chips, and even garbage. A majority of the material in plants is cellulose. Cellulose degradation occurs in

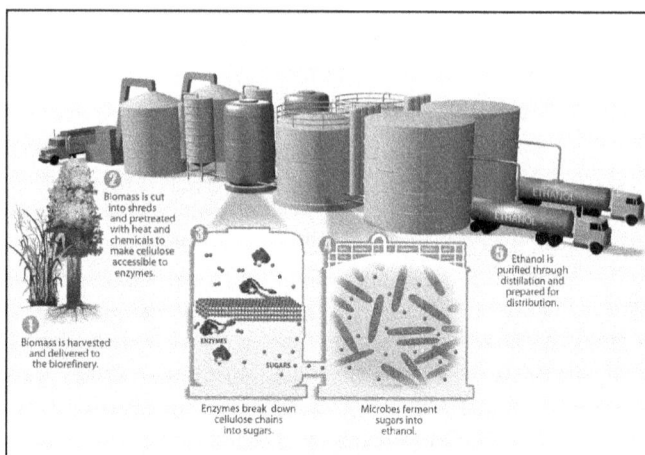

Figure 6.2 *How cellulosic ethanol is made*
Source: Genome Management Information System, Oak Ridge National
Laboratory

nature, but slowly. The challenge is to increase the efficiency of cellulase, an enzyme that breaks down cellulose, and to improve the yield of cellulose from biological sources.

Subsidies on corn production and tax exemptions for non-petroleum fuels have been instrumental in enabling entrants to produce and market biofuels. The situation for biofuels is analogous to the early years of penicillin production. The basic scientific principles are known, a strong market need exists, but a better method is needed to enable cost-effective large-scale production.

PLASTIC

The world's first modern biorefinery, a Cargill-Dow project, went online in Blair, Nebraska in 2002. The plant is the product of a joint venture established in 1997 between Cargill and the Dow Chemical Company to commercialize polylactic acid (PLA) under the brand name NatureWorks. Dow pulled out of the venture in 2004 acknowledging significant long-term potential, but dissatisfaction with short-term profitability.

PLA is made by fermenting the sugar in corn (other high-sugar feedstocks are also amenable to this process) into lactic acid molecules, which are then linked to form polylactic acid. PLA can be used to make a wide array of products, including plastic cups and containers, wrappers, and polyester textiles. Furthermore, PLA is biodegradable, requires 65 percent less energy to produce than conventional plastics, and can reduce fossil fuel use in plastic manufacture by up to 80 percent.

Bio-based plastics have the added benefit of being naturally biodegradable, reducing the environmental impact of their use. They face resistance due to their higher costs, the need to re-engineer downstream manufacturing and utilization processes in some cases, and reduced suitability in harsh environments. Despite these hurdles, they are finding strong adoption in consumer-facing applications such as disposable packaging for food containers.

OTHER BIO-BASED PRODUCTS

As described in the section *Biofuels* above, the original source of petroleum products is actually biological. Accordingly, the potential exists for biotechnology innovations to replace petroleum products in many manufacturing processes.

Vitamin B2 is used as a supplement in animal feed to keep animals healthy and fit. In 1990 BASF developed an innovative fermentation method to replace the traditional eight-step chemical process used for vitamin B2 production, using *Ashbya gossypii* fungus with a one-step fermentation. This fermentation process reduces costs by up to 40 percent and reduces environmental impact by 40 percent. The fermentation process has several other advantages over chemical synthesis. BASF has realized a 95 percent reduction in waste, reduced energy usage due to lower reaction temperatures, and a 60 percent reduction in the resources required.

Another product improved by biotechnology is propanediol. Propanediol is a clear colorless liquid with applications in deicing, as an engine coolant, in adhesives and coatings, and as an additive in cosmetics and shampoo. Traditionally produced from petroleum feedstocks at high temperatures, a joint venture of Dupont and Tate & Lyle uses a proprietary fermentation method to produce a biologically-derived version of propanediol named Bio-PDO for industrial and consumer applications. The benefits of Bio-PDO over conventional alternatives are reduced production energy requirements, low toxicity, and biodegradability, and improved heat stability and reduced corrosion when used as anti-freeze. Bio-PDO is also replacing petroleum sources in the manufacture of Dupont's Sorono plastic.

DETERGENTS

Detergent enzymes represent the broadest application of enzymes. Detergent enzymes improve household laundry, dishwashing, and industrial washing applications by improving cleaning performance, reducing washing times, reducing energy consumption by lowering wash temperatures, and even rejuvenating clothes.

The most common enzymes used are proteases and amylases, which respectively remove stains and soils based on proteins and starches. Other enzymes with applications as detergent adjuncts include lipases to digest fat or oil based stains, peroxidases to inhibit dye transfer, and cellulases to prevent pilling on cotton clothes.

These innovations allow a reduction in the use of numerous environmentally-damaging chemicals, including phosphates and bleaches.

MINING

Microorganisms are used worldwide in mining processes to oxidize and leach metals. Other applications are as alternatives to harsh chemicals to remove metals from industrial wastewater streams. The primary bacteria employed are *Thiobacillus ferrooxidans*, *Leptospirillum ferrooxidans,* and thermophilic (high temperature) bacteria to leach metals such as copper and gold from sulfide minerals. Some of the advantages of bioleaching over conventional roasters, smelters, and pressure autoclaves are that construction time is shorter, no noxious gases or toxic effluents are produced, environmental permit and reporting processes are simpler, and safety is increased due to processing at or near ambient temperatures and pressures.

Another method under development is the use of plants

Table 6.2 *Selected industrial enzymes*

Enzyme type	Function and utility
amylase	Decomposes simple sugars. Applications in textiles, laundry and dishwashing, biofuels, and paper production.
cellulase	Decomposes cellulose into simpler sugars. Applications in biofuel production, laundry, and paper processing.
lipase	Decomposes fats. Applications in laundry and surface cleaning, food processing, leather processing, and pharmaceuticals.
protease	Decomposes proteins. Applications in laundry, leather processing, baking.
xylanase	Degrades plant cell walls. Applications in paper production, biofuels, food production, and textiles.

to mine sparse deposits of valuable minerals. Plants with enhanced abilities to sequester heavy metals in soil can extract sparse deposits of gold or other valuable minerals, which can then be recovered by simply harvesting and incinerating the plants.

OIL WELL COMPLETION

Microbial enhanced oil recovery (MEOR) is the use of microorganisms to retrieve recalcitrant oil from existing wells, maximizing petroleum production of an oil reservoir. MEOR employs the inoculation of selected natural bacterial strains into oil wells to decrease the viscosity of thick oil deposits and ease oil flow, or to produce gases such as carbon dioxide to propel oil out of the well.

BIOREMEDIATION

Bioremediation is the application of biotechnology for environmental reclamation. Some of the processes described above have applications in bioremediation. Relative to existing alternatives, the use of plants, microorganisms, and their by-products to sequester pollutants, or to degrade them into relatively benign compounds, can be a safer, cheaper, and faster method to clean the environment.

Unresolved questions regarding the release of genetically

Box

Using bacteria to make snow

S nowmax is an ice-nucleating protein derived from naturally-occurring *Pseudomonas syringae* bacteria. It is hypothesized that the natural purpose of this ice-nucleating protein is as part of a long-distance dispersion strategy of *Pseudomonas syringae*, which is a plant pathogen. The bacteria are able to survive for long periods in aerial suspension. The ice-nucleating proteins help drop the bacteria out of aerial suspension, by way of rain or snow, enabling them to infect plants.

Specially-designed aeration guns are used to spray water mixed with Snowmax on ski slopes. Snowmax increases the number of nucleation centers in water droplets from these aeration guns, improving snow making efficiency and also enabling snow making at higher temperatures, ultimately saving ski resorts money and improving the quality of ski slopes. Snowmax is produced by growing *Pseudomonas syringae* in a fermentation vessel and extracting the protein using filtration processes. The product is irradiated prior to shipping to ensure that live bacteria, which are not harmful to humans, are not released.

modified organisms into the environment motivate the search for more natural techniques. Fortunately, there are few natural materials that at least one naturally-ocurring microorganism cannot use as a nutrient. Given appropriate conditions, even synthetic compounds are subject to microbial metabolism. By searching for organisms already feeding on pollutants, either in natural environments or at polluted sites, it is possible to develop non-transgenic bioremediation systems with applications ranging from treatment of oil spills to reclamation of contaminated soil and water.

RED BIOTECHNOLOGY: MEDICAL APPLICATIONS

MONOCLONAL ANTIBODIES

Antibodies are natural components of human and other immune systems that recognize unfamiliar material such as infectious bacteria and cancerous cells and help eliminate them. While our natural complement of antibodies is generally very effective at recognizing and prompting the destruction of infectious microorganisms and cancerous cells, threats are sometimes missed. Monoclonal antibodies are a category of biotechnology-derived drugs that are designed to act and look like naturally occurring antibodies and may directly treat diseases or condition a patient's own immune system to launch a highly specific attack on infections or diseased tissues. They are designated "monoclonal" because they are produced as large batches of identical molecules. Georges Köhler and César Milstein received the 1984 Nobel Prize in Physiology or Medicine for their description of a technique to produce monoclonal antibodies. They shared the prize with Niels Jerne, who described the development and control of the immune system.

In the 1980s the first antibody trials saw early experimental therapies rendered inactive by the liver, or activating patients' own immune systems to raise antibodies against the foreign therapeutic antibodies, resulting in increased illness. The rejection of these initial non-human antibodies can be attributed to the primary purpose of the immune system: to repel foreign bodies. The use of modified versions of animal antibodies, humanized antibodies, and fully human antibodies led to the development of monoclonal antibody therapies that are safe and effective.

Genentech's Rituxan was the first monoclonal antibody to be approved for cancer treatment in the United States. Rituxan works by binding to specific types of cancer cells and triggering the immune system to destroy them. Another Genentech product, Herceptin, is targeted at growth factors that are directly implicated in approximately 20 percent of breast cancers (see Box *Personalized medicine and drug sales* later in this chapter).

Antibody-based drugs can enjoy years of strong sales with minimal competition, even after patent expiration, because of the difficulty of demonstrating equivalence of antibodies produced by a second party. This challenge, however, motivates competitors to produce improved antibodies rather than simply producing competing antibodies. This means that when an innovator's antibody is challenged by a new entrant, the innovator is likely to lose a greater portion of market share than they might if the competing drug were merely equivalent to the original.

RNA INTERFERENCE

RNA interference therapies, sometimes referred to as anti-sense therapies, block gene expression. Summarizing information presented in Chapter 3, mRNA is a molecule that transfers information from genes to the protein synthesis machinery within cells. The goal of RNA interference is to intercept an mRNA message before it is translated into protein. Andrew Fire and Craig Mello shared the 2006 Nobel Prize in Physiology or Medicine for their discovery of gene silencing by RNA interference.

It is important to recognize that RNA interference is a subtractive solution. Interference cannot directly replace, amplify, or add a gene function. It can only inhibit a gene function (although inhibiting an inhibitory gene can indirectly increase the expression of a second gene). RNA interference has the potential to treat a range of diseases including cancer, autoimmune disorders, and infectious diseases.

A successful RNA interference therapy must enable entry of interfering molecules into cells to permit a therapeutic effect, prevent degradation of interfering molecules before they can act, and ensure specificity so that essential functions are not disrupted. The relative ease of controlling these issues in laboratory settings means that many RNA interference therapies that look promising in pre-clinical development are likely to face complications in therapeutic settings. Isis Pharmaceuticals' Vitravene, the first RNA interference drug, overcomes delivery issues through direct injection into the target tissue.

GENE THERAPY

Gene therapy uses genes to treat disease. Techniques for gene therapy include replacement of defective genes, and supplementation with therapeutic genes. For diseases caused by an absent or defective copy of a specific gene, supplementation of that gene can potentially cure the disease. The technical challenges of gene therapy include targeting appropriate cells and tissues, ensuring gene transfer, controlling gene expression, and satisfying safety concerns.

While most genetic deficits require gene expression in specific cell types, some diseases can be cured by expression of specific genes in a variety of cells. Blood-related deficiencies, such as the lack of clotting factors in hemophilia, can potentially be cured by enabling the cells lining blood vessels to produce the necessary clotting factors. A caution for untargeted therapy is that certain cell types may suffer complications from expression of inappropriate genes. Another potential complication is that a patient's immune system may reject an introduced gene product and the cells producing it, leading to destruction of healthy tissue.

Regulation of quantity and duration of gene expression is also important. While diseases such as cystic fibrosis and hemophilia require persistent expression and may be cured even with low expression, other diseases such as diabetes require tightly regulated and coordinated gene expression. For some genetic diseases, irreparable damage occurs early in life. For example, cystic fibrosis leads to lung damage during childhood. It is important to intervene and treat such diseases before permanent damage is sustained.

An early success for gene therapy was witnessed in a 1990 trial when two girls with a genetic deficit causing severe immunodeficiency were given infusions of their own immune system cells. These cells were genetically engineered to contain a working version of their missing gene. Following regular monthly administration, the girls developed active immune systems that allowed them to remain healthy for more than 10 years.

The first marketed gene therapy product was approved in China in 2003. Shenzen SiBiono's Genicide is targeted at head and neck squamous cell carcinoma, a highly lethal cancer with an annual incidence of 300,000 people in China. The drug uses a benign viral vector to deliver *p53*, a gene implicated in controlling cell growth; many tumors contain defective *p53* or fail to express sufficient quantities of the protein.

Despite its early promise, advances in gene therapy have been hampered by variability in the safety and effectiveness of trials, sometimes with fatal consequences.

DIAGNOSTIC TESTS

In addition to treating diseases, biotechnology has also made it easier to detect and diagnose medical conditions. A quantum leap past traditional techniques that require correlation of numerous symptoms to develop a diagnosis, biotechnology enables the direct detection of biological processes. In addition to refining symptom-based diagnoses, it is also possible to make determinations at earlier stages. Screening for pregnancy and cancer are examples of diagnoses that have increased in reliability and sensitivity as a result of biotechnology.

For diseases where symptoms are usually noticed past the point where treatment is most effective, diagnostic tests can save lives by enabling at-risk individuals to monitor their health prior to onset of disease. Individuals with genetic predispositions to specific cancers can be alerted to their increased likelihood of disease and can engage in preventative activities and regular screenings, potentially avoiding disease progression or allowing early intervention.

A secondary benefit of diagnostic tests is that they can enable individuals to avoid costly, dangerous, or unnecessary procedures (see Box *Personalized medicine and drug sales* later in this chapter). For example, use of aspirin to prevent colorectal cancer may be less cost-effective than regular screening. Citing a cost of nearly $150,000 per year of life saved by preventative use of aspirin, factoring in the costs of the drug and treatment of side effects, versus a $30,000 cost of screening per year of life saved, a recent study concluded that screening was preferable to aspirin use.[3]

On initial examination, diagnostics may seem to be a good market-entry objective for start-ups seeking to develop an initial revenue stream. Diagnostics are relatively cheaper to develop than therapeutics and it is also relatively easier to gain regulatory approval for them. However, because drugs serve more pressing needs than diagnostics they can be assured of relatively greater sales and greater profits due to decreased

3 Ladabaum, U., *et al.*, Aspirin as an adjunct to screening for prevention of sporadic colorectal cancer. *Annals of Internal Medicine*, 2001. 135(9):769-781

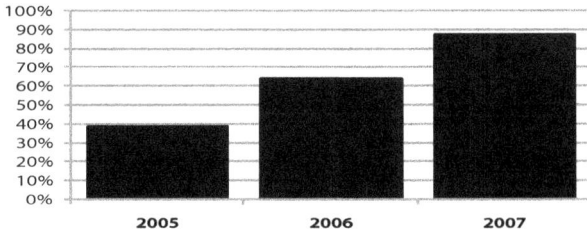

Figure 6.3 *U.S. babies born in states mandating genetic disorder testing*
Source: March of Dimes

price elasticity.

PERSONAL GENETIC PROFILING

As information about genetic markers increases and DNA profiling and sequencing costs decrease, numerous companies are entering the personal genetic profiling space. These companies offer to profile or sequence an individual's DNA and identify predispositions to disease, with the objective of enabling their customers to take proactive approaches to prevent disease onset.

Aside from fraudulent operations seeking to sell vitamin mixtures of dubious benefit to consumers, the very science of disease predisposition mapping is in its infancy. Many diseases have strong environmental and genetic components, and the influences of these factors need to be understood for genetic profiling to be useful. Faced with a poorly-documented association of a genetic sequence with a disease, or conflicting reports of genetic predisposition, testing companies and patients are likely to assume a stronger connection than is warranted by the data.

PERSONALIZED MEDICINE

Personalized medicine involves the application of technologies such as functional genomics (see Chapter 5) to tailor therapies to the patients most likely to benefit from them. It is estimated that most commonly used drugs are effective in only 30–60 percent of patients with a given disease. A subset of these patients may suffer severe side effects. There are two causes for this difference in response. First, most diseases have myriad causes. They tend to be defined by symptoms, but each distinct cause may respond best to a different treatment. Second, people are different. Differences in liver metabolic enzymes, for example, can determine if patients are unlikely to respond to a drug or if they will suffer severe side effects—see the section *Pharmacogenetics* in Chapter 5 for more details.

Aligning treatments with the patients most likely to benefit from them holds the potential to improve the effectiveness of medical intervention, while reducing healthcare costs and dangers. The key to realizing personalized medicine is alignment of a diagnostic test with a therapeutic intervention. A well-paired test, such as the test for Her-2 overexpression which indicates that Genentech's Herceptin is the preferred drug, can have a strong positive impact on sales. In other cases, personalized medicine may mean fewer sales.

While fewer sales may represent a barrier to implementation of personalized medicine in some cases, it can also be a benefit. Using profiling in clinical trials can speed approval. It is estimated that the time saved in Herceptin's clinical trials netted Genentech between $1.2 and $1.5 billion.

TISSUE ENGINEERING

Tissue engineering is the production of natural or synthetic organs and tissues which may be implanted as fully functional units, or as tissue which undergoes further development following implantation to perform necessary functions. The first engineered tissues were skin equivalents used to treat burn victims, and structural scaffolds to replace heart valves, arteries, and bones. Alternative treatments for tissue and organ failure include transplantation from donors, surgical repair, artificial prostheses, mechanical devices, and in a few cases, drug therapy. Tissue engineering has the potential to provide an alterna-

---

Box

Personalized medicine and drug sales

Personalized medicine has great potential to improve the safety and efficacy of drugs by targeting therapies to those most likely to benefit from them and excluding patients who are susceptible to deleterious side effects. While patients generally stand to benefit from personalized medicine, biotechnology companies may benefit or suffer based on whether screening methods are used to identify, or to exclude, patients. The contrasting examples of Herceptin and Aczone demonstrate how personalized medicine can benefit or hurt drug sales.

Herceptin

Genentech's Herceptin is a monoclonal antibody directed at the Her-2 cell receptor. Overexpression of Her-2 is implicated in approximately 20 percent of breast cancers. It is estimated that without a test for Her-2 overexpression, Genentech would have needed to perform clinical trials on 2,200 patients for ten years in order to demonstrate the efficacy of Herceptin.[1] Utilizing a test for over-expression of Her-2 to segregate patients, Genentech was able to demonstrate Herceptin's ability to safely increase survival times by 50 percent using only 469 patients in less than two years.

Because the test for Her-2 overexpression is tied to the mode of action of Herceptin, patients are able to avoid many of the unnecessary side effects associated with ineffective medicines and may benefit from early prescription of the drug most likely to effectively treat their tumors, while Genentech benefits from preferred prescription to patients who test positive for Her-2 overexpression.

Aczone

QLT's Aczone is a topical drug used to treat acne. In the course of clinical trials it was found that people with a blood disorder called G6PD deficiency have a higher risk of developing anemia with Aczone; roughly 1.4 percent of patients in clinical trials had this disorder. The potential for anemia among patients with G6PD deficiency taking Aczone spurred the FDA to require that patients be tested for the enzyme deficiency prior to being prescribed Aczone.

Unlike Herceptin, where the diagnostic test is optional prior to prescription and identifies the target population, Aczone prescription requires prior testing to exclude patients likely to suffer deleterious side effects. The impact of this excluding screen is that while patients are protected from adverse reactions, QLT suffers a significant barrier to prescription.

1   Tansey, B. Genentech a big believer in diagnostics, *San Francisco Chronicle,* May 17, 2004. p. B-1.

tive or complement to these treatments, potentially with fewer side effects and a greater ability to treat major damage.

While some cell types and tissues are amenable to production in liquid media or on solid surfaces, a challenge for production of more complex tissues and organs is the development of appropriate scaffolds to model growth and methods to direct local differentiation of tissues. Large organs must be perfused by blood vessels to allow for oxygenation, delivery of nutrients, and removal of waste. An alternative to laboratory production of implantable organs is the use of stem cells, which can potentially be coaxed to repair tissues upon injection into patients.

Tissue engineering also faces significant commercial challenges. Stem cells and xenotransplantation offer alternative methods to serve many of the same markets as tissue engineering. The history of tissue engineering also serves as an example of how quickly fortunes can turn in biotechnology. Despite research and development expenditures of $4.5 billion, as of 2002 none of the tissue engineering products on the market were profitable.[4] By 2007, the sector consisted of 50 profitable firms which had treated over a million patients and were generating $1.3 billion in annual sales.[5]

STEM CELLS

Stem cells can repair damaged organs and tissues and even have the potential to produce entire organs in laboratory settings for use as human replacement parts. Understanding and controlling the ability of stem cells to repair organs holds the potential to eliminate the need for transplantation and tissue engineering. Unlike most of the cells in human adults, stem cells are able to differentiate into other cell types. Degenerative diseases such as Alzheimer's disease and Parkinson's disease, as well as diseases marked by cell injury or malfunction, such as stroke, heart attack, cancer, and spinal cord injury, are all candidates for stem cell therapy.

4 Lysaght, M.J., Hazlehurst, A.L. Tissue engineering: the end of the beginning. *Tissue Engineering*, 2004. 10(2):309-320.

5 Lysaght, M.J., Jaklenec, A., Deweerd, E. Great expectations: Private sector activity in tissue engineering, regenerative medicine, and stem cell therapeutics. *Tissue Engineering Part A*, February 1, 2008. 14(2):305-315.

Stem cells can be harvested from remnants of fertility treatments, placentas, umbilical cords, and from adult tissues. Stem cells from embryonic sources such as discarded fertilized eggs from fertility treatments carry a significant ethical and political burden. Additionally, there are restrictions on the use of federal funding for embryonic stem cell research in the United States. Stem cells extracted from adult sources such as bone marrow do not share the same ethical and funding issues as embryonic stem cells, but it is clear that they have different properties as well.

There are three categories of stem cells, distinguished by their ability to differentiate into other cell types. Totipotent cells can grow into an entire organism. Embryonic stem cells are unique in having this property. Pluripotent cells cannot grow into a whole organism, but they are able to differentiate into the various cell types of the body and potentially form organs. Multipotent (sometimes called unipotent) cells can only form certain types of cells such as blood cells or bone cells. Embryonic stem cells are pluripotent and multipotent by virtue of their totipentcy, while adult stem cells may be pluripotent or multipotent depending on their source.

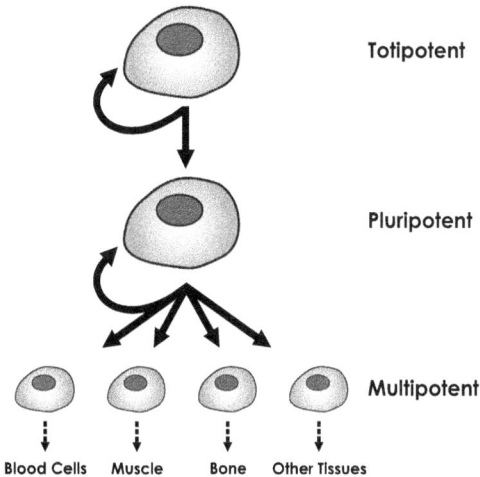

Figure 6.4 *Stem cell types*
Modified from National Institutes of Health

A market for leukemia treatment with adult stem cells already exists. A growing body of research also indicates that adult stem cells can cure hearts and other organs that have suffered trauma. Interestingly, implantation of embryonic stem cells has been implicated in cardiac arrhythmias and cancer development, suggesting that adult stem cells may be better suited for these applications.

Stem cells are also being explored as a vehicle to deliver genes to specific tissues in the body for gene therapy or cancer treatment. The greatest technical challenge facing stem cell researchers is elucidating the factors that activate stem cells to form specific kinds of tissue. Isolating and purifying sufficient quantities of stem cells for clinical use presents an additional challenge. Extensive patenting also makes freedom to operate an important consideration.

XENOTRANSPLANTATION

Shortages of human organs available for transplant and disease considerations have prompted the search for alternative sources of organs. Xenotransplantation is the transplantation of organs from any other species into humans. The similarity of human and pig organs has led most xenotransplantation research to be focused on pigs. Immune rejection and the potential for novel infections to spread into human populations are the most pressing challenges facing xenotransplantation.

Xenotransplantation need not involve implantation of whole organs. Studies have shown survival of fetal pig neural cells when administered to patients with Huntington's or Parkinson's disease. The potential also exists to coax such cells to perform needed functions. It is also possible to use animal organs for *ex vivo* treatments; use without implantation. In 1997, a patient with liver failure had his blood perfused through a liver from a transgenic pig raised by Nextran (Baxter donated Nextran to the Mayo Clinic in 2003), keeping him alive for over six hours until a liver donor could be found.[6]

Rejection may occur by multiple mechanisms with fast or slow time profiles. Methods for dealing with rejection range

6 Stolberg, S.G. Could this pig save your life? *New York Times,* October 3, 1999.

from administering immunosuppressive agents to removing immunoreactive elements through genetic engineering of donor animals. Aside from rejection, a significant challenge for xenotransplantation is the prospect of novel infections being introduced through the transplant recipient into the human population. One way to partially alleviate this risk is to genetically engineer animals to remove known risk factors.

Glossary

SCIENCE

Absorption, Distribution, Metabolism, Excretion and Toxicology (ADMET): An element of pre-clinical and clinical trials used to measure the effects of a drug on animal and human physiology.

Amino acid: Building block of proteins. Proteins consist of amino acids linked end-to-end. There are 20 different amino acid molecules that make up proteins. The DNA sequence that codes for a gene dictates the order of amino acids in a given protein.

Antibiotic: A chemical substance that can kill or inhibit the growth of a microorganism.

Antibody: Immune system protein produced by humans and higher animals to recognize and neutralize bacteria, viruses, cancerous cells, and other foreign compounds.

Antisense: A natural or synthetic DNA or RNA molecule that specifically binds with messenger RNA to selectively inhibit expression of a single gene.

Applied research: Aimed at gaining knowledge or understanding to determine the means by which a specific recognized need may be met. Applied research builds upon the discoveries of basic research to enable commercialization.

***Bacillus thuringiensis*:** A naturally occurring bacteria that produces Bt toxin, a protein that is toxic to certain kinds of insects. The Bt toxin gene has been genetically engineered into corn and cotton plants to reduce the need for chemical pesticides.

Base: A key component of DNA and RNA molecules. Four different bases are found in DNA: adenine (A), cytosine (C), guanine (G) and thymine (T). In RNA, uracil (U) substitutes for thymine.

Basic research: Aimed at gaining more comprehensive knowledge or understanding of the subject under study, without specific applications in mind.

Biofuel: Fuels such as ethanol and diesel produced from sugars, vegetable oils, or other organic matter using biotechnology methods.

Bioinformatics: The application of information technology to manage and analyze the vast amounts of data generated from biological research.

Bioleaching: The use of plants to extract heavy metals from soils.

Bioremediation: The use of biological systems, usually microorganisms, to decompose or sequester toxic and unwanted substances in the environment.

Biotechnology: The application of molecular biology for useful purposes.

Blue biotechnology: A seldom-used term referring to marine and aquatic applications of biotechnology.

Chromosome: The DNA-protein complexes that contain all the genes in a cell.

Cloning: The process of making an identical copy of something. Often used in reference to copying animals, it may also refer to creating copies of DNA fragments, individual cells, or plants.

Codon: A sequence of three DNA or RNA bases that specifies an amino acid in the synthesis of a protein.

Combinatorial chemistry: A product discovery technique that uses robotics and parallel chemical reactions to generate and screen as many as several million molecules with similar structures in order to find chemical molecules with desired properties.

Cytochrome p450: A set of enzymes involved in chemical modification and degradation of chemicals including drugs and other foreign compounds.

Data mining: Using computers to analyze masses of information to discover trends and patterns.

Diagnostic: A product used for the diagnosis of a disease or medical condition.

DNA (deoxyribonucleic acid): The primary source of genetic information in cells. DNA is comprised of nucleotides and is composed of two strands wound around each other, called the double helix.

DNA fingerprinting: A DNA analysis method that measures genetic variation among individuals. This technology is often used as a forensic tool to detect differences or similarities in blood and tissue samples at crime scenes.

DNA sequencing: The process of determining the exact order of bases in a segment of DNA.

Double-blind: An experimental protocol whereby neither the experimental subjects nor the administrators know whether a drug or placebo is being administered. Double-blind protocols are used to eliminate bias.

Drug delivery: The process by which a formulated drug is administered to the patient.

Drug development: The process of taking a lead compound, demonstrating it to be safe and effective for use in humans, and preparing it for commercial-scale manufacture.

Enzyme: A functional protein that catalyzes (speeds up) a chemical reaction. Enzymes control the rate of naturally occurring metabolic processes such as those necessary for growth and reproduction.

***Escherichia coli (E. coli)*:** A common gut bacteria that is a workhorse and model organism for molecular biology.

Excipient: An inactive ingredient (there are no absolutely inert excipients) added to a drug to give it a pill form or otherwise aid in delivery.

Expression: A highly specific process in which a gene is switched on at a certain time and its encoded protein is synthesized, resulting in the manifestation of a characteristic that is specified by a gene. Genetic predispositions to disease arise when a person carries the gene for a disease but it is not expressed.

False negative: An experimental outcome that incorrectly yields a negative result. False negatives can complicate disease diagnosis.

False positive: An experimental outcome that incorrectly yields a positive result. False positives can frustrate assessing the performance of lead compounds.

Fermentation: Technically the process of breaking complex organic substances into simpler ones, such as conversion of sugars into alcohols, acetone, or lactic acid. Also refers to any large-scale cultivation of microbes or other single cells (e.g., for drug production).

Functional genomics: The use of biological experiments and genetic correlations to establish what each gene does, how it is regulated, and how it interacts with other genes.

Functional foods: Foods containing compounds with beneficial health effects beyond those provided by the basic nutrients, minerals, and vitamins.

Gene: The fundamental unit of heredity, a segment of DNA which encodes a defined biochemical function. Some genes direct the synthesis of proteins, while others have regulatory functions.

Gene expression: The production of a gene product—generally defined as the synthesis of an encoded protein.

Gene splicing: Splicing a gene from one segment of DNA into another. Commonly used to insert foreign genes into bacteria for analysis, or to insert foreign genes into bacteria or other organisms for genetic modification or to produce and harvest large quantities of specific proteins.

Gene therapy: The replacement of a defective gene in a person or organism suffering from a genetic disease.

Genetic code: The language in which DNA's instructions are written. The genetic code consists of triplets of nucleotides (codons), with each triplet corresponding to one amino acid in a protein structure, or a signal to start or stop protein production.

Genetic disorder: A condition or mutation that results from one or more defective genes.

Genetic engineering: The manipulation of genes to create heritable changes in biological organisms and products that are useful to people, living things, or the environment.

Genetic predisposition: A susceptibility to disease that is related to a genetic condition, which may or may not result in actual development of the disease.

Genetic screening: The use of a specific biological test to screen for inherited diseases or medical conditions.

Genome: The sum of an organism's genes.

Genomics: The study of genes and their function.

Good manufacturing practice (GMP): Guidelines ensuring the quality and purity of chemical products that are intended for use in pharmaceutical applications, and controls ensuring that methods and facilities used for production, processing, packaging, and storage result in drugs with consistent and sufficient quality, purity, and activity.

Gray biotechnology: A seldom used term for industrial applications of biotechnology. More commonly referred to as white biotechnology.

Green biotechnology: The use of biotechnology for agricultural applications.

Human Genome Project: The international research effort which identified and located the full sequence of bases in the human genome.

Incidence: measure of the rate of new occurrences of a disease or condition in a population.

Immune system: The cells, biological substances (such as antibodies), and cellular activities that work together to recognize foreign substances and provide resistance to disease.

In silico **(in computer):** Computer-based predictions that can complement *in vitro* and *in vivo* procedures.

In vitro **(in glass):** Experimental procedures carried out in testtubes, beakers, etc.

In vivo **(in the living body):** Experimental procedures carried out on living cell lines or in living animals.

Lead compound: In pre-clinical development and clinical trials, a potential drug being tested for safety and efficacy.

Liposome: An artificial membrane. Can be used to encapsulate drugs and aid in drug delivery.

Microarray: A tool that permits the identification of DNA samples and examination of gene expression in individual tissues and different conditions.

Monoclonal antibody: A synthetic immune system protein that recognizes a single target. Polyclonal antibodies recognize several related targets.

Molecular evolution: The process of making discrete changes in genes to improve the functional characteristics of proteins and enzymes.

Molecular farming: Using biotechnology to produce useful products from domesticated plants and animals.

mRNA (messenger RNA): A ribonucleic acid molecule that transmits genetic information from DNA to the protein synthesis machinery in cells, where it directs protein synthesis.

Mutant: A cell or organism harboring one or more mutated genes.

Mutation: A change in the base sequence of a gene that results in it not performing its normal task.

Nanotechnology: A technology field focusing on materials at sizes measured in billionths of a meter.

Nucleotide: One of the structural components, or building blocks, of DNA and RNA. A nucleotide consists of a base plus one molecule of sugar and phosphoric acid.

Oncogenic: Viruses, chemicals, genes, proteins, etc. that cause the formation of tumors.

Pathogen: A disease-causing organism.

Personalized medicine: The practice of medicine in which therapies are developed for and directed at the patients most likely to benefit from them.

Pharmacogenetics: Examination of the differences in drug response between individuals—one drug, many genomes.

Pharmacogenomics: Examination of differences in how one person responds to different drugs—many drugs, one genome.

Pharming: The process of farming genetically engineered animals and plants to produce drugs.

Placebo: A mock-treatment used in single-blind or double-blind experiments to eliminate bias from experiment subjects or administrators, respectively.

Platform technology: A technique or tool that enables a range of scientific investigations. Examples include combinatorial chemistry for producing novel compounds, microarrays for gene expression analysis, and bioinformatics programs for data assembly and analysis.

Polymerase Chain Reaction (PCR): A method to produce sufficient DNA for analysis from a very small amount of DNA.

Prevalence: measure of how commonly a disease or condition occurs in a population.

Prion: A naturally occurring protein that can be converted into a disease-causing form. Prion diseases can be transmitted in the absence of DNA or RNA.

Promoter: A DNA sequence preceding a gene that contains regulatory sequences influencing the expression of the gene.

Proof-of-principle: Demonstration of the commercial potential of a discovery or invention.

Protein: A long-chain molecule comprised of amino acids that folds into a complex three-dimensional structure. The type and order of the amino acids in a protein is specified by the nucleotide sequence of the gene that codes for the protein. The structure of a protein determines its function.

Proteomics: The study of the protein profile of each cell type, protein differences between healthy and diseased states, and the function of, and interaction among, proteins.

Rational drug design: Using the known three-dimensional structure of a molecule, usually a protein, to design a drug that will bind have a therapeutic effect on it.

Recombinant DNA: The DNA formed by combining segments of DNA from different sources.

Red biotechnology: The use of biotechnology for therapeutic applications.

Reformulation: Altering an established drug's formulation or delivery method to yield improvements in safety or efficacy.

Repurposing: Finding new indications for approved drugs.

Restriction enzyme: A protein that cuts DNA molecules at specific sites, dictated by the nucleotide sequence.

Retrovirus: A type of virus that reproduces by converting RNA into DNA.

Single Nucleotide Polymorphism (SNP): A single base difference in the sequence of a gene which alters the structure and function of the gene product.

RNA (ribonucleic acid): A nucleic acid, similar to DNA, which has roles in gene expression.

RNA interference: Using antisense techniques to selectively inhibit expression of a gene.

Stem cell: An undifferentiated cell that can multiply and become any sort of cell in the body.

Telomere: The tip of a chromosome. Telomeres are involved in the replication and stability of chromosomes.

Tissue engineering: The production of natural or synthetic organs and tissues that can be implanted as fully functional units or may develop to perform necessary functions following implantation.

Transcription: The synthesis of an mRNA molecule as a copy of a gene. In gene expression, transcription precedes translation.

Translation: The synthesis of a protein based on the nucleotide sequence of an mRNA molecule, which corresponds to the sequence of a gene.

Transgenic: An organism with one or more genes that have been transferred to it from another organism.

Vaccine: A preparation of either whole disease-causing organisms (killed or weakened) or parts of such organisms, used to confer immunity against the disease that the organisms cause. Vaccine preparations can be natural, synthetic, or derived by recombinant DNA technology.

White biotechnology: The use of biotechnology for industrial applications.

X-ray crystallography: An essential technique for determining the three-dimensional structure of biological molecules.

Xenotransplantation: Transplanting a foreign tissue into another species.

LEGAL

Claim: A comprehensive and precise description that defines the scope of an invention.

Compulsory license: A license in which a government forces the holder of a patent to grant use to the state or others. Authorized under WTO provisions to enable countries to produce generic versions of patented drugs in the event of a health crisis.

Continuation: A filing, while a patent is active, which contains additions or changes to the previous claims.

Copyright: The exclusive legal right to publish, perform, display, or distribute an original work.

Divisional patent: A patent that covers the same specification as a previous (parent) patent, but claims a different invention.

***Ex parte*:** A legal proceeding where only one party is represented. Patent prosecution is an *ex parte* procedure.

Experimental use: The practice of a patented invention solely with intention of experimentation or perfection of the invention.

Evergreening: The practice of launching new formulations, combinations, delivery methods, and indications for drugs facing patent expiration to effectively increase the duration of patent-protected sales.

Filing date: The date on which a complete patent application is received by the Patent and Trademark Office.

Freedom to operate: The absence of intellectual property and regulatory impediments (which may require patent license and passage of enabling laws) to commercialization.

***Inter partes* reexamination:** A method by which third parties challenge the validity of a patent on the grounds of prior art publication without resorting to litigation.

Interference: When two or more patent applications or issued patents claim the same invention.

License: An agreement whereby one party gains access to another's technology (e.g. a patent license).

Non-disclosure agreement (NDA): An agreement, common between companies and their contractors and partners, which allows a company to share protected information while preventing its release.

Office action: A formal response by a patent examiner regarding a patent application or amendment.

Patent: A description of an invention. Patents contain one or more claims that describe the subject matter covered in sufficient detail to permit skilled experts to practice an invention, and grant the right to exclude others from practicing an invention.

Patent agent: An individual with technical training who is capable of representing an inventor in patent prosecution.

Patent attorney: An individual with legal training in patent law who is capable of representing an inventor in patent prosecution and litigation.

Patent pool: An agreement between two or more patent owners to license one or more of their patents to one another or third parties.

Patent term adjustment: Provisions to adjust patent term to provide restoration for U.S. patent and trademark office delays.

Patent term extension: Provisions to extend patent term to account for time spent waiting for FDA approval.

Prior art: Public knowledge that exists in a field; all previously issued patents, publications, public announcements, or knowledge that bear on the invention claimed in a patent application.

Prosecution: The process by which an inventor engages with the patent office to obtain a patent and determine the scope of its claims.

Provisional patent application: A preliminary patent application filed without a formal patent claim, oath or declaration, or any prior art statement. It provides the means to establish an early effective filing date in a subsequent non-provisional patent application and allows the term "Patent Pending" to be applied.

Reach-through claim: A patent claim to rights to royalties from, or rights to use, drugs or other physical or intellectual property produced using the patent

Submarine patent: A patent that emerges after it has unknowingly been infringed upon.

Trade secret: Knowledge and information that is not generally known to the industry. Examples include customer lists, business plans, and manufacturing methods.

Trademark: A registered name, word, symbol, or device identifying a company's products or services.

REGULATORY

Abbreviated New Drug Application (ANDA): A simplified submission permitted for a generic version of an approved drug.

Accelerated approval: A process to make products for life threatening diseases available on the market prior to formal demonstration of benefit. Uses surrogate markers—indirect measures of efficacy—and requires continued testing to confirm efficacy.

Action letter: An official FDA communication that informs the sponsor of an NDA or BLA of a decision by the agency. An approval letter allows commercial marketing of the product.

Authorized generic: Drugs produced by branded companies and marketed under a private label to compete with other generic drugs.

Bayh-Dole Act: Provides the statutory basis and framework for federal technology transfer activities, including patenting and licensing federally funded inventions to commercial ventures.

Bioequivalence: Demonstration that a generic drug has the same chemical and biological properties as its pioneer counterpart.

Biologic: Medicine made by biological processes rather than by chemical synthesis or extraction. Biologics typify biotechnology-derived drugs. Contrast with small-molecule drugs.

Biologics License Application (BLA): Application filed with the FDA Center for Biologics Evaluation and Research (CDER) for approval to market a biologic drug.

Biosimilar: A generic biologic drug that is "similar but not identical" to a pioneer drug.

Brand-name drug: The original, often patented, version of a drug. Contrast with generic drugs.

Clinical pharmacology study: Clinical trial designed to determine the absorption, distribution, metabolism, elimination, and toxicity (ADMET) of a drug.

Clinical trial: A human study designed to measure the safety and efficacy of a new drug.

Current good manufacturing practices (cGMP): Regulatory practices to ensure safety and consistency of manufacturing processes.

Exclusivity: A temporary FDA-granted monopoly, distinct from patent or other intellectual property protection. Exclusivity may be granted for developing drugs for rare diseases, novel drugs, conducting pediatric clinical trials, or successfully challenging invalid patents.

Fast track: A process for interacting with the FDA during drug development, intended for drugs to treat serious or life threatening conditions that demonstrate the potential to address an unmet medical need.

First-in-man study: Phase I trial primarily concerned with establishing the safety of a compound.

Follow-on biologic: An FDA term for a biologic drug that is similar to an existing biologic.

Generic drug: The version of an approved drug produced by a competitor after a pioneer firm's patents expire.

Hatch-Waxman safe harbor: A research-use exemption stemming from the Hatch-Waxman Act which exempts from infringement the use of patented inventions in preparation for submitting drug applications.

Hatch-Waxman Act: Contains provisions to foster the development of generic drugs and support pioneer drug development.

Indication: A use for which a specific drug is approved by the FDA.

Institutional Review Board (IRB): An independent committee of scientists, physicians, and lay people that oversees clinical trials.

Investigational New Drug (IND): An application to pursue clinical trials with an experimental drug that has passed pre-clinical development.

March-in rights: A stipulation of the Bayh-Dole Act enabling the government to request and potentially require issuance of a license to a patent, which was developed with federal funding, to another party.

Named Patient Program: European compassionate use program, enabling limited distribution of drugs prior to approval.

New Drug Application (NDA): Application filed with the FDA Center for Drug Evaluation and Research (CDER) for approval to market a small-molecule drug.

Off-label use: Use of a drug not in accordance with FDA-approved uses or drug labeling. Physicians are free to prescribe drugs for off-label uses.

Orange Book: Also known as *Approved Drug Products with Therapeutic Equivalence Evaluations*, the *Orange Book* contains detailed information on all approved drugs and must list all extant patents.

Orphan Drug: A drug that treats a disease affecting fewer than 200,000 Americans or for which there is no reasonable expectation that the cost of research and development will be recovered from sales in the United States. The Orphan Drug Act provides special incentives for producers of orphan drugs.

OTC-switch: The process of gaining approval to sell a drug over the counter, which may grant 3 years exclusivity for the over-the-counter market.

Over the counter (OTC): Selling a drug without a prescription. Requires evidence that patients can self-diagnose and use the drug safely without physician supervision.

Phase I: Clinical trial designed primarily to determine the safety of an experimental drug.

Phase II: Clinical trial that evaluates an experimental drug's safety, assesses side effects, and establishes dosage guidelines.

Phase III: Clinical trial designed to assess the safety and effectiveness of an experimental drug. Success in Phase III trials can lead to marketing approval.

Phase IV: Post-approval clinical trials used to monitor safety and efficacy or examine additional applications of drugs.

Pioneer (brand-name) drug: The patented version of a drug. Contrast with generic drugs, the competing versions produced when pioneer patents expire.

Pre-clinical studies: Studies that test a drug on animals and non-human test systems. Safety information from such studies is used to support an investigational new drug application (IND).

Reverse payment: A payment from a branded drug company to a generic drug company to delay launch of a generic drug.

Salami slicing: Filing for multiple orphan drug designations on the same drug.

Small-molecule drug: A drug produced using defined chemical synthesis or extraction. Contrast with biologics, drugs produced by biological processes.

Surrogate marker: An indirect measure of effectiveness, such as a laboratory test or tumor shrinkage, used to show a strong potential for effectiveness in accelerated drug approval.

COMMERCIAL

Accredited investor: A type of investor, largely defined by their wealth, permitted to invest in high-risk investments.

Acquisition: Appropriation of the controlling interests of one company by another.

Alliance: Agreement between two or more companies to cooperate in some way.

Angel investor: Wealthy individual who personally provides startup capital to very young companies to help them grow.

Barrier to entry: A condition that makes it difficult for competitors to enter the market (e.g., patent, trademark, high up-front capital requirements).

Blockbuster: Drug with $1 billion or more in sales.

Board of directors: A group legally charged with the responsibility to protect the interests of a company and its shareholders.

Bootstrap: Starting a business with little or no external funding.

Bridge loan: A short-term, high-interest, loan provided to companies in dire need of cash.

Burn rate: The rate at which an unprofitable company is going through its available cash reserves.

Business model: A description of a company's purpose, commercial offerings, strategies, organizational structure, operational processes, etc. Often confused with business plan, below.

Business plan: A formal statement of a company's goals and the plan for reaching those goals.

Comparable: A valuation technique based on analogy to similar companies or products.

Competitive advantage: An advantage that a firm has relative to competing firms; may be in the form of intellectual property, expertise, partnerships, assets, etc.

Controlling interest: Ownership of more than 50 percent of a company's voting shares.

Convertible: Securities (usually bonds or preferred shares) that can be converted into common stock.

Cooperative research and development agreement (CRADA): An agreement enabling federally funded laboratories to perform for-profit contract work for commercial firms.

Corporate inversion: Formation of a parent corporation of a U.S. company in a country with little or no corporation tax, and structuring a U.S. subsidiary to manage U.S. sales. This scheme enables foreign sales to be taxed in local markets only, and not be taxed in the U.S.

Cross-licensing: An agreement in which two or more firms with competing and similar technologies strike a deal to reduce the need for legal actions to clarify who is to profit from applications of the technology.

Dilution: The decrease in relative ownership among existing investors as additional shares are issued.

Discounted cash flow (DCF): A valuation technique that attempts to consider future events and determine a present value for a product or project based on the variety of outcomes.

Discovery rights: Selling only research findings while retaining rights to knowledge discovered in the course of research and development.

Down round: A financing event in which a company is valued lower than it was previously.

Due diligence: The process by which research is conducted to determine the value of an investment, licensing agreement, merger, or other similar activity.

Earnings strippings: A potential result from a corporate inversion where payments from a U.S. subsidiary to a foreign parent are used as deductions against U.S. taxes.

Elevator pitch: A short summary—typically less than two-minutes—used to quickly describe a business to investors

Dumb money: Funding from investors who cannot provide additional benefits such as guidance or networking.

Equity dilution: The dilution of the equity stakes of founders and early investors by subsequent investments.

Equity investment: An investment purchasing partial ownership of a company.

Exit: The means by which investors gain a return on their investment, commonly through sale of shares in public markets or acquisition by another company.

Free cash flow: The amount of cash available to a company after all expenses have been paid.

Friends and family: A term for investments which often help start a company, and are typically made by unaccredited investors.

Incubator: A facility offering space and shared services and facilities to early-stage companies.

Initial Public Offering (IPO): The initial sale of shares of a private company in public markets, turning it into a publicly-traded company.

Institutional support : Esteem granted to companies by their affiliation with highly regarded partners, financiers, and other affiliates.

Intellectual property: Intangible assets such as patents, trade secrets, trade names, etc.

License: An agreement to grant rights to a patent or tangible subject.

Market segmentation: The division of a market into distinct groups of buyers or decision makers.

Medicaid: Government-subsidized healthcare coverage for individuals with low incomes and limited resources.

Medical tourism: Travelling for medical treatment. Often motivated by cost-savings, local waiting lists and expertise availability, or local regulatory restrictions.

Medicare: Government-subsidized healthcare for individuals 65 years of age and older, some disabled people under 65 years of age, and people with permanent kidney failure.

Merger: The formal combination of two companies into one entity. Often used to refer to acquisitions. A merger can be distinguished from consolidation, in which a new separate entity is created.

Mezzanine funding: Funding that generally leads to liquidity (IPO or merger) or commercial launch and eventual profitability.

Milestone: The completion of a specified phase in product development. Investors and alliance partners may use milestones to establish a timeline for incremental investments or payments.

Offshoring: The relocation of business processes from one country to another.

Options-pricing: A valuation technique that analyzes the value of discrete operational paths.

Outsourcing: The execution and management of selected operations by outside parties.

Parallel trade: The trade of products between countries without permission of the intellectual property owner. Often used to capitalize on inter-country price differences.

Pharmacoeconomics: Study of the cost-benefit ratios of drugs.

PIPE (Private Investment in Public Equity): Purchase of discounted shares in a public company in which payment goes directly to the company rather than to existing shareholders.

Preferred stock: A convertible offering that cannot be sold until it is converted to common stock.

Price elasticity: A measure of the change in demand resulting from a change in price of a product or service. Low price elasticity indicates little change in demand; high elasticity indicates a relatively large change in demand.

Price/earnings (P/E) ratio: A rudimentary technique to determine the value of a company, or relative value of several companies, by comparing the share price to annual earnings.

Private equity: In contrast to owning shares in a public company, private equity is ownership in a private company.

Proxy fight: A process by which shareholders can vote for corporate changes. May be used to appoint new directors or replace senior management.

Royalty: The payment of a percentage of sales as compensation to product developers, patent licensors, or even investors.

Ratchet: An anti-dilution provision where an investor is granted additional shares of stock without charge if the company later sells the shares at a lower price.

Return on Investment (ROI): Profit (or loss) on an investment, often expressed as a percentage.

Reverse merger: The merger of a private company with a "shell" company, rendering the private company public.

SBIR (Small Business Innovation Research): A funding program that encourages small business to explore their technological potential and provides the incentive to profit from its commercialization.

Scientific Advisory Board (SAB): A group of esteemed scientists and business professionals, independent from management, which provides objective feedback and guidance on a company's progress and goals.

Seed financing: Capital furnished to prove the feasibility of a concept or invention.

Secondary offering: A public or private share offering subsequent to an initial public offering.

Series A/B/C/D: Venture funding stages that fund product development and early commercial launch activities.

Shell: A public company with few or no assets that may be the remnant of a bankruptcy or asset sale. Used in reverse mergers to enable a private company to become public.

Smart money: Funding from investors who are able to contribute guidance, networking, or other benefits.

Special purpose acquisition company (SPAC): A public company formed with the intent of engaging in a reverse merger with a private company.

Spin-off: Separating a smaller unit from an established company, permitting each company to retain focus while shielding the parent from risk and granting the spin-off the administrative benefits of small size.

STTR (Small Business Technology Transfer): A funding program that encourages public/private sector partnership in order to develop new technologies and profit from their commercialization.

Targeted marketing: The alignment of marketing efforts with the benefits sought by individual market segments.

Tax arbitrage: Location of specific operations in countries and regions with favorable tax treatment for those operations.

Technology transfer: The transfer of discoveries made by basic research institutions, such as universities and government laboratories, to the commercial sector for development into useful products and services.

Venture capital: Money invested by venture capitalists in startup companies in exchange for equity.

Venture capitalist: An individual who invests in start-up companies with the intent of making a large return on investment.

Virtual company: Firms that outsource all or most of the elements of research, development, and marketing.

Index

About the Author

Yali Friedman is managing editor of the *Journal of Commercial Biotechnology* and serves on the science advisory board of Chakra Biotech and the editorial advisory boards of the *Biotechnology Journal, Journal of Medical Marketing* and *Open Biotechnology Journal.* He regularly guest-lectures for biotechnology education programs, teaching classes on the business of biotechnology, and has written and given talks on diverse topics such as biotechnology entrepreneurship, strategies to cope with a lack of management talent and capital when developing companies outside of established hubs, and new paradigms in technology-based economic development.

Yali also has a long history in biotechnology media, having created a *Forbes* "Best of the Web"-rated web site on the biotechnology industry for a NY Times company and managed it for many years. His other projects include the Student Guide to DNA Based Computers, sponsored by FUJI Television, BiotechBlog.com, and DrugPatentWatch.com, a pharmaceutical industry competitive intelligence service.

Yali can be contacted at *info@thinkbiotech.com.*

Related titles from Logos Press
www.logos-press.com/books

BUILDING BIOTECHNOLOGY
An expanded version of this guide, this is the definitive primer on the business of biotechnology
Softcover: 978-09734676-6-6
Hardcover: 978-09734676-5-9

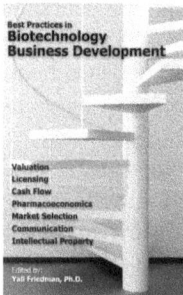

BEST PRACTICES IN BIOTECHNOLOGY BUSINESS DEVELOPMENT
Eleven chapters from biotechnology industry experts
ISBN: 978-09734676-0-4

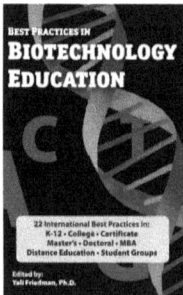

BEST PRACTICES IN BIOTECHNOLOGY EDUCATION
22 chapters on programs from 5 countries
ISBN: 978-09734676-7-3